Kuwait Soil Taxonomy

A typical sandy soil of Kuwait "Entisols" covering 30 percent of the surveyed area

Shabbir A. Shahid · Samira A. S. Omar

Kuwait Soil Taxonomy

Springer

Shabbir A. Shahid
Desert Agriculture and Ecosystems Program
Environment and Life Sciences Research Center
Kuwait Institute for Scientific Research
Safat, Kuwait

Samira A. S. Omar
Kuwait Institute for Scientific Research
Safat, Kuwait

ISBN 978-3-030-95296-9 ISBN 978-3-030-95297-6 (eBook)
https://doi.org/10.1007/978-3-030-95297-6

Photos of plants are the courtesy of Dr. Samira A. S. Omar

This Springer imprint is published by the registered company Springer Nature Switzerland AG
The registered company address is: Gewerbestrasse 11, 6330 Cham, Switzerland

Rhanterium epapposum is the national plant of Kuwait. Rhanterium grows well in calcigypsids, torripsammts and petrocalcids soils

Foreword

Rationale uses of soils require their potential to be scientifically assessed. This is essential as misuse of soils can compromise their quality without achieving the targeted benefits. In this context, national soil classification becomes more important in the current situation, where, in 2050, it is likely that the agrifood sector will face a momentous challenge to produce high–quality and nutritious food to feed 9 billion people, while dealing with the impact of climate change on natural resources. Keeping in mind that 95% of our food is directly or indirectly coming from land-based agriculture, and to feed 9 billion people by 2050, the UN Food and Agriculture Organization (UNFAO) expects an increase of 60% food globally and 100% in developing countries, where it is also indicated that currently 33% of global soils are moderately to highly degraded, and 21% of the irrigated lands are affected by soil salinization. However, if the soils are properly understood and managed sustainably, it could be possible to produce up to 58% more food.

Considering the importance of the soils of Kuwait for agriculture, food security and in the provision of ecosystem services, the Kuwait Institute for Scientific Research (KISR) in collaboration with the Public Authority for Agriculture Affairs and Fish Resources (PAAFR) led the efforts to update the national soil map in 1995 and completed in 1999. The soil map was updated by systematizing and processing information through dedicated soil surveys at the national level (1:100,000 scale) and 200,000 ha area having the potential of irrigated agriculture (1:25,000 scale), using the latest soil survey and classification standards of the United States Department of Agriculture-Natural Resources Conservation Service (USDA-NRCS). We thank the contribution of USDA experts in conducting QA/QC on the soil survey of the State of Kuwait and appreciate the team efforts at KISR and PAAFR for completing the survey successfully.

The publication of Kuwait Soil Taxonomy is a continuity of earlier efforts to understand the quality of soils for diversified uses for national development. The book will be a guide to identify soil classes at unknown places for science-based land use planning and informed decisions for the implementation of specific projects. I believe the book is a definitive and essential reference for potential stakeholders including but not necessarily limited to students, studying soils as part of their curriculum, earth or environmental sciences, as well as professional soil scientists and land use planners. I congratulate the authors for this unique publication, which will be a reference to use for the next at least 3–4 decades. Last but not least, I would like to thank Dr. Curtis Monger, National Leader, Soil Survey Standards USDA-NRCS National Soil Survey Center, for his approval to extract material from USDA-NRCS publications relevant to Kuwait Soil Taxonomy.

Dr. Afaf Al-Nasser
Acting Executive Director
Environment and Life Sciences Research Center
Kuwait Institute for Scientific Research
Safat, Kuwait

Arnebia decumbens annual herb that is one of the most common of the desert flowers that blooms early in spring

Preface

The soils are an integral part of the terrestrial life to provide food, construction material and ecosystem services. We all know that for centuries, humanity and soils have lived one aside from the other. The rise and fall of civilization have been linked to the use and abuse of soil and water resources. We live on the planet Earth which gives us soils to use and live on. Where the soils are a finite resource and non-renewable. Their sustainable use for diversified functions, such as ecosystem service and food security, is inevitable. The misuse of soil resources reduces resource quality for positive services; therefore, the maintenance of soil quality is essential. The sustainable use of soil resources without compromising the quality suggests its quality is to be assessed and maintained if not improved. This requires the soils in any country to be used rationally based on the potential for a specific use, which is only possible if the soils are described and classified using the internationally recognized soil classification system suiting the country's environment and the soils. Once the soils are classified and soil and diversified thematic maps are published, these become guidelines for informed decisions on land use planning, national development and policy implications.

National soil classification system is established to help the nation to predict soil behavior and to provide a common language for soil scientists. National soil survey is a systematic study of the soil of a country, which includes soil classification using internationally recognized procedures and mapping of soil properties and the distribution of various soil map units. The soil map units are characterized in a systematic way that useful interpretations of the soils can be made for potential uses and response to changes in management. Considering the importance of soil mapping in land use planning and national development, a multi-million-dollar soil survey of Kuwait project was completed jointly by KISR and PAAFR through an international professional contractor AACM International, Adelaide Australia, during 1995–1999. The soil survey of Kuwait has revealed the heterogeneous landscapes including coastal and inland sabkhas, sand sheet and sand dunes and hence presenting soil diversity in terms of physical, chemical, mineralogical and fertility properties. During the reconnaissance Soil Survey of Kuwait, of 12 Soil Orders covering the entire globe, in Kuwait two soil orders *(Aridisols and Entisols),* seven suborders, eight great groups, 14 subgroups and 24 soil families are identified, and in the semi-detailed soil survey, 39 soil series have been identified. Overall, *Entisols* covered 30% and *Aridisols* covered 70% of the surveyed area of Kuwait. These valuable reports and maps are essential resources; however, there was a gap of readily available guidelines to further classify soils in the deserts of Kuwait for location-specific projects. The Kuwait Soil Taxonomy has filled this gap and is a useful guide to provide a mechanism to update the previous soil surveys conducted in Kuwait and will facilitate the soil correlations from new soil surveys in Kuwait. We believe that the "Kuwait Soil Taxonomy" will also enhance the utilization of national soil information and associated maps for many years to come. In this book, the terms not found in the twelfth edition of US soil taxonomy are added in the Kuwait soil taxonomy by including the criteria and classes for identifying "salidic" subgroup within the great groups of *Petrogypsid*s, *Argigypsids*, *Haplogypsids* and *Torriorthents*.

The main aim to publish *Kuwait Soil Taxonomy* is to provide easy-to-understand guide for keying out soils of Kuwait into different soil classes and access to associated laboratory soil procedures. The soil survey of Kuwait was completed by using the ninth edition of US Keys to Soil Taxonomy which is upgraded to the twelfth edition of the USDA-NRCS and Soil Survey Manual published in 2017. The Kuwait Soil Taxonomy is a combination of information from Soil Survey of Kuwait reports predominantly extracted and correlated with the twelfth edition of the USDA-NRCS Keys to Soil Taxonomy, and sections relevant to the soils found in Kuwait are included. It provides a useful tool, not only for soil analysts and specialists, but also for decision-makers who seek to promote innovative processes for the development of the country in general, and the agricultural sector in particular.

Safat, Kuwait

Shabbir A. Shahid, Ph.D.

Samira A. S. Omar, Ph.D.

Scrophularia deserti grows on hard rocky ground

Acknowledgements

We greatly acknowledge the vision of the management of both the Kuwait Institute for Scientific Research (KISR) and the Public Authority for Agriculture Affairs and Fish Resources (PAAFR), who took the initiative to jointly complete the Soil Survey and Associated Activities of Kuwait project during 1995–1999. The project was implemented through an internationally reputed professional contractor Australian Agriculture Consultant Management (AACM) Adelaide, Australia. A large team both from KISR and PAAFR joined the AACM to implement various technical tasks of the project. We wish to pay special tribute to KISR Director General, Project Leader of the Soil Survey Project and Dr. Afaf Al-Nasser who was the Manager of the Aridland Agriculture and Greenery Department (AAGD) for their interest in the project from the inception to the completion. Dr. Shabbir Shahid had the opportunity to work with the soil survey team in the capacity of technical coordination to assure the project Terms of Reference are fully complied and the quality of the final reports and the associated maps is met. The report and maps generated through this eminent project are precious resources for broad land use planning and national development. However, Dr. Samira Omar strongly believes there was still a gap of readily available guidelines for those who wish to further explore soil classification in the deserts of Kuwait, which necessitated the development of the *Kuwait Soil Taxonomy* book that will be used as a ready reference by the students at Kuwait University, researchers, professionals, landscape contractors and other potential users as a guide to classify soils in the area of their interest for many years to come. The photos of plants from the deserts of Kuwait are generously provided by Dr. Samira A. S. Omar.

We wish to thank Dr. Curtis Monger, National Leader—Soil Survey Standards USDA-NRCS National Soil Survey Center—for providing the approval to access USDA-NRCS publications and extract relevant parts for the Kuwait Soil Taxonomy. The cooperation of USDA-NRCS during the implementation of Soil Survey Activities and in the finalization of Kuwait Soil Taxonomy is highly acknowledged.

The approval of this book for publication by Sheikh Mohammed Yousif Saud Al-Sabah Director General Public Authority for Agriculture Affairs and Fish Resources (PAAFR), Dr. Mane Alsudairawi Acting Director General KISR, Dr. Afaf Al-Nasser (OD/DD-ELSRC), Dr. Sameer Al-Zenki (DD/STD-ELSRC) and Majda K. Suleiman (Program Manager, Desert Agriculture and Ecosystems Program) is highly acknowledged and appreciated.

Shabbir A. Shahid, Ph.D.

Samira A. S. Omar, Ph.D.

Preamble

Sheikh Mohammed Yousif Saud Al-Sabah

Globally 95% of food is directly or indirectly produced on soils. To have optimum food production from soils and sustaining soil health for long-term services should be taken as a national priority. This requires an understanding of national soils and their rational uses based on their productive capacities for specific projects.

Soils are important for the provision of multiple ecosystem services and the basis for food, feed, fuel and fiber production. Considering the importance of soils of Kuwait for agriculture and food production, Kuwait's Public Authority for Agriculture Affairs and Fish Resources (PAAFR) jointly with Kuwait Institute for Scientific Research (KISR) through international consultant AACM International Australia has completed the Soil Survey of Kuwait in the year 1999. After this completion, it was essential to compile the Kuwait Soil Taxonomy book to enhance the use and application of soil information and maps. PAAFR has been engaged with KISR for decades to enhance applied agricultural research in Kuwait. The joint aim is to promote local food production based on scientific evidence, that we can achieve economically to reduce dependency on food import, and to avoid disruption in the food supply chain as was the case during the COVID-19 pandemic in many countries. The purpose of the book *Kuwait Soil Taxonomy* is to bridge the knowledge gap about the soils of Kuwait and to help guide potential stakeholders to classify soils and to establish their potentials for national development projects. This is because some soils may be marginal for one use and vital for other uses. The book has come at a time when the world is making significant efforts to double food production by 2030 as a commitment to United Nations-Sustainable Development Goal 2 (End hunger). The book is considered a way forward, well ahead of 2030 to achieve this target in Kuwait. It is believed that the multiple stakeholders of this book, including students, academia, researchers, professionals, decision-makers and land use planners will give this book due consideration in shaping national development in general, and agriculture and food security in specific.

I am very pleased to learn that finally the important task is accomplished. Further, it is a great honor to learn that Kuwait is among the few countries in the world to publish such a unique book. I congratulate the authors for this significant and unique achievement and fully support such professional efforts in the future which are of mutual interests to both PAAFR and KISR for the development of Kuwait.

Sheikh Mohammed Yousif Saud Al-Sabah
Director General
Public Authority for Agriculture Affairs
and Fish Resources (PAAFR)
State of Kuwait

Cistanche tubulosa growing in sandy desert soil

National Leader Soil Survey Standards

Curtis Monger

The Kuwait Soil Taxonomy is an excellent source of information for use by diversified stakeholders who wish to classify soils in Kuwait to locate suitable sites for new projects and infrastructure development. I found the book with excellent contents is written by experienced scientists actively involved in the implementation of Soil Survey of Kuwait Project. The book is published on the latest USDA-NRCS soil classification standards, relevant to the soils of Kuwait. It establishes soil classification standards of Kuwait that can be used for many years to come. The book was peer-reviewed by USDA-NRCS Scientist Dr. Craig Ditzler, an experienced soil taxonomist. The authors are congratulated for this significant achievement which will be of interest not only to agricultural scientists but also to the decision-makers in Kuwait.

Curtis Monger, Ph.D.
Professor Emeritus
New Mexico State University
Las Cruces, New Mexico and Former National Leader
Soil Survey Standards, USDA-NRCS (retired)

Phragmites australis growing in wet land soil-marshes

International Reviewer Kuwait Soil Taxonomy

Craig Ditzler

This book, *Kuwait Soil Taxonomy*, is an important resource for understanding the land resources of Kuwait. A soil survey of the state of Kuwait utilizing the same standards as those used in the USA was published in 1999. Perhaps the most important aspect of the soil survey is that it not only identifies the kinds of soil occurring in the country, but the maps provide a geographic context for where each of the soils occurs. This is very important for those wanting to understand the soils at a particular location. The Kuwait soil survey utilized the then-current USDA system of soil classification to classify the soils. Since 1999, the USDA Soil Taxonomy has been improved in a number of ways. This includes some changes to the way arid soils such as those of Kuwait are classified. This publication, while based on the original Kuwaiti soil survey, updates the taxonomic information to reflect the most current revisions to the system. Improvements include the kinds and definitions of diagnostic horizons and features, standard conventions for describing soils in the field, new classes identified at the family level of classification and updated taxonomic keys for determining the correct classification of the soil based on its' properties. All of these updates and more are included in this new publication.

In addition to providing the most current information on soil classification as it applies to Kuwaiti soils, this publication provides some basic information briefly describing how the USDA classification compares to some other systems used around the world. Another important section provides information about the various procedures for collecting samples in the field and measuring various key properties of the soil in the laboratory. It is very important to not only know what the measured value of a key soil property is, but also to know how it was determined.

Kuwait Soil Taxonomy is an important contribution for anyone wanting to understand the land resources of Kuwait. I congratulate the authors for an important job very well done.

Craig Ditzler, Ph.D.
Retired USDA National Leader for Soil Survey Standards
Lincoln, Nebraska, USA

Contents

Rostaria pumila grass growing in sandy desert

About the Authors

Dr. Shabbir A. Shahid was embraced with prestigious *Sir William Roberts* award to pursue a Ph.D. degree in Soil Science specialization in Soil Micromorphology of Salt-affected Soils at the University of Bangor Wales UK, completed in 1989. He has over 40 years experience as a soil scientist in Pakistan, UK, Australia, United Arab Emirates and Kuwait. Currently, Dr. Shahid is Research Scientist, Desert Agriculture and Ecosystems Program, ELSRC-KISR. He was a technical coordinator in multi-million-dollar national soil surveys of the State of Kuwait and Abu Dhabi Emirate and developed the soil survey action plan for the Northern Emirates of UAE and the Republic of Mauritania. Dr. Shahid, with his co-associates, discovered anhydrite soil which is formally added in the twelfth edition of US Keys to Soil Taxonomy as a diagnostic horizon, mineralogy class and subgroups in the Salids suborder of the order Aridisols. He is also the Principal author of United Arab Emirates Keys to Soil Taxonomy published by Springer. In addition, Dr. Shahid is a creator and co-founder of the Emirates Soil Museum launched in 2016 at the International Center for Biosaline Agriculture, Dubai United Arab Emirates.

He is a prolific author of over 160 scientific papers published in peer-reviewed scientific journals, book chapters, conference proceedings and newsletters. As Editor/Co-editor/Principal author, he

Erodium glaucophyllum a perennial herb grows on rocky terrain on shallow sandy soil

published seven books by professional publisher Springer. At the time of book preparation, Dr. Shahid $\mathbf{R^G}$ score was 26.44, h-index 20 with 1,610 citations, 1,880 research interests and over 127,000 reads globally. His $\mathbf{R^G}$ score of 26.44 is higher than 82.5% of all Research Gate members' scores.

Dr. Samira A. S. Omar was appointed by the Kuwait Council of Ministers as the Director General of KISR from 2016–2021. She was responsible for the Science, Technology and Innovation (STI) development and promotion in the country. Currently, Dr. Omar is Principal Research Scientist conducting full-time Research & Development (R&D) at KISR. She has led many projects to conserve biodiversity, restore ecosystems and promote sustainable agricultural development in Kuwait. Dr. Samira also led the Soil Survey for the State of Kuwait Project from 1995–1999 and the Kuwait Environmental Remediation Program (KERP) for the United Nations Compensation Commission (UNCC) from 2011–2014. Dr. Samira is affiliated with many international organizations, such as the International Union for Conservation of Nature (IUCN) and the Society for Ecological Restoration (SER). Dr. Omar is a Research Fellow at The World Academy of Sciences (TWAS) and has received many awards for her scientific achievements. Recently, she received the Decoration of OSI (Order of the Star of Italy) at the very high rank of Grand Officer for her collaboration with the Government of Italy in Research & Development (R&D) in 2021. She was also honored with the Theodore M. Sperry Award 2019 by SER in September 2019. Dr. Samira holds a Ph.D. degree in wild land resource science from the University of California, Berkeley. She is a prolific author of many peer-reviewed research papers, conference proceedings and book chapters. She published many books on the vegetation of Kuwait, protected areas and remediation of environmental pollution.

Tamarix aucheriana growing in the saline coastal flat

Acronyms and Abbreviations

AACM	Australian Agriculture Consultant Management
AAGD	Aridland Agriculture and Greenery Department
AAS	Atomic Absorption Spectrophotometer
AASHTO	American Association of State Highway and Transportation Officials
°C	Degree centigrade
CEC	Cation-Exchange-Capacity
COVID 19	Coronavirus Disease 2019
DAEP	Desert Agriculture and Ecosystems Program
dS/m	deci Siemens per meter
DTA	Differential Thermal Analysis
EAD	Environment Agency Abu Dhabi
EC	Electrical Conductivity
ECe	Electrical conductivity of soil saturation extract
ELSRC	Environment and Life Sciences Research Center
ESP	Exchangeable Sodium Percentage
°F	Degree Fahrenheit
FAO	Food and Agriculture Organization
HCl	Hydrochloric acid
ICP	Inductively Coupled Plasma
IRA	Infrared Analysis
IUSS	International Union of Soil Sciences
KISR	Kuwait Institute for Scientific Research
kPa	Kilo Pascal
MAF	Ministry of Agriculture and Fisheries
MAW	Ministry of Agriculture and Water
MoEW	Ministry of Environment and Water

PAAFR	Public Authority for Agriculture Affairs and Fish Resources
SAR	Sodium Adsorption Ratio
SI System	International System of Units
TWAS	The World Academy of Sciences
UNESCO	United Nations Educational, Scientific and Cultural Organization
USA	United States of America
USDA-NRCS	United States Department of Agriculture-Natural Resources Conservation Service
WRB	World Reference Base
XRDA	X-Ray Diffraction Analysis
XRF	X-Ray Fluorescence

List of Figures

List of Tables

Water erosion near Jal Az zor escarpment

Soil Classification Systems and Kuwait Soil Taxonomy Hierarchy

1

Abstract

Rationale uses of soils require the potential of soils to be established through systematic soil classification at national levels. This is essential as misuse of soils (e.g., irrigation with saline water, soil pollution with excessive fertilizer use etc.) compromise their soil quality without achieving the targeted benefits. Sustainable use of soils and to feed 10 billion peoples by 2050, it is essential to use soils rationally without compromising soil quality for sustainable intensification. Where soil resources are considered to be the most important factor in meeting that challenge, given that almost 95% of our food is produced directly or indirectly in the soil. Considering the above facts, it is essential to innovate and practice agriculture on our soils that produce more and at the same time conserve and sustain soil resources for years to come to serve next generations, this is possible when we understand our soils properly through established national and international soil classification systems. This also helps sharing the established agricultural technology for adoption, from similar environments and soils. Few countries have established their national soil classification systems (Australia, Canada, China, England and Wales, France, New Zealand, Norway, Russia, South Africa and United Arab Emirates) which are briefly described in this chapter. Similar soil classification system for Kuwait has not yet been established and hence

S. A. Shahid and S. A. S. Omar, *Kuwait Soil Taxonomy*,
https://doi.org/10.1007/978-3-030-95297-6_1

become the focus of this publication. In other countries international systems (FAO; World Reference Base-WRB or Soil Taxonomy) were adapted to local edaphic conditions.

Keywords
Hierarchy · Soil classification · World Reference Base · Soil taxonomy · Kuwait

1.1 Introduction

National soil inventory is important in the current situation, where, in 2050 it is likely that agrifood sector will face a momentous challenge to produce high quality and nutritious food to feed 9 billion peoples (Charles et al. 2010), that is 3 billion more mouths to feed than there were in 2010, while dealing with the impact of climate change on natural resources. Soil classification deals with the systematic categorization of soils based on distinguishing characteristics as well as criteria that dictate choices in use. It separates soil into classes or groups each having similar characteristics and potentially similar behavior. There exists number of soil classification systems. According to Food and Agriculture Organization-FAO, Soil classification concerns the grouping of soils with a similar range of properties (chemical, physical, mineralogical and biological) into units that can be geo-referenced and mapped. In the soil survey of Kuwait, we used United States Department of Agriculture-Natural Resources Conservation Service (USDA-NRCS) soil classification system, which has found widespread international acceptance particularly in countries in Latin America and Asia. The principles that were developed by Soil Taxonomy were taken up by World Reference Base (WRB) and the FAO Legend to set international standards.

1.2 History of Soil Classification Systems

The earlier USDA soil classification (Baldwin et al. 1938) divided soils into three orders and was focused on the environment and the soil forming factors to classify soils in zonal, azonal and intrazonal soils. A later development focused on the processes occurring in the soil itself. These processes were roughly characterized by soil properties. A good example of the latter approach is the French classification system (CPCS 1967). Modern soil classification started with the publication of the 7th Approximation of the USDA Soil Taxonomy (Soil Survey Staff 1975), where precisely defined and quantified soil properties as such, or in combination, were used to define "diagnostic soil horizons". This has undergone several revisions. It is a hierarchical classification that tries to group similar soils into increasingly general categories. It was designed to support soil survey in the USA, however, it tries to classify all World soils into 12 soil orders (Soil Survey Staff 2014).

1.2.1 Soil Classification Systems

Currently a number of countries have established the national soil classification systems, that often focused on the specific characteristics of the soils within the national territory. Other countries where local soil classification do not exist, they have used international systems (FAO, WRB or US Soil Taxonomy) by adapting to local edaphic conditions. These soil classification standards have been used in more than 75 countries of the world. In the Gulf Cooperation Council Countries US Soil Taxonomy standards have been used, such as the Kingdom of Saudi Arabia (MAW 1985), Sultanate of Oman (MAF 1990), State of Kuwait (KISR 1999a, b), State of Qatar (Scheibert et al. 2005), Dubai emirate (Dubai Municipality 2005), Abu Dhabi emirate (EAD 2009a, b), and Northern Emirates of UAE (EAD-MOEW 2012). Later based on the Abu Dhabi emirate soil classification results, a comparative soil classification using USDA and FAO system for Abu Dhabi coastal area is discussed by Abdelfattah and Shahid (2007).

Other systems in use are Soil Classification for England and Wales (Avery 1980), the Soil Map Legend of the World (FAO-UNESCO 1990), World Reference Base for Soil Resources (WRB) (IUSS Working Group WRB 2015), the Canadian System of Soil Classification (Soil Classification Working Group 1998), the Australian Soil Classification (Isbell 2016), the New Zealand Soil Classification (Hewitt 1992), and Russian Soil Classification System (Shishov et al. 2004). These soil classification systems are briefly described below.

1.2.1.1 Australian Soil Classification System (2016)

It presents the knowledge of Australian soils for those who use the land. The Isbell in 1992 observed two classification schemes widely used prior to 1996, these are, (i) The Handbook of Australian soils (Stace et al. 1968) and the Factual Key (Northcote 1979). The advantages and disadvantages of these two Classification schemes have been discussed by Moore et al. (1983).

In the past a vast amount of soils data has been accumulated and compiled in 2016 edition (Isbell 2016) with a number of guiding principles. One major difference of Australian Soil Classification with US Soil Taxonomy was no depth restrictions, such as the arbitrary lower limit of 2 m used in US Soil Taxonomy. The Australian Soil Taxonomy soil classification system is a highly-developed with detailed hierarchical classification where special emphasis is given on highly-weathered soils and soils of arid and semi-arid regions and based on a large volume of high-quality soil data.

1.2.1.2 Canadian Soil Classification System

Prior to 1955, the Canadian soil testing was similar to the methods used in the United States. It is only in 1955, a Canadian soil classification was introduced. Compared to soil classification systems used globally, the Canadian System of *Soil Classification* (Haynes 1998) is more closely related to USA system, but differ in many ways. It is a hierarchical system where the classes are conceptual and generalization of properties of real bodies of soil. Soil taxa are defined based on observable and measurable soil properties reflecting soil *genesis* and environmental factors. The development of the system has progressed with the increasing knowledge of the soils of Canada obtained through pedological surveys carried out over 80-year period.

1.2.1.3 Chinese Soil Taxonomy

In China soil classification has a long history, however, the modern work was initiated in 1930's, where taxonomic classification was not started until 1980's. Finally, an English version of Chinese Soil Taxonomy published in 2001 (Feng 2001) based on globally accepted diagnostic horizons and characteristics, and using over 400 related papers. Later Shi et al. (2004) presented reference benchmarks relating to great groups of genetic soil classification of China with soil taxonomy. Whereas cross-reference system for translating between genetic soil classification of China and soil taxonomy was published by Shi et al. (2006), and with WRB at different scales by Shi et al. (2010).

1.2.1.4 England and Wales Soil Classification System

The soils of England and Wales are classified based on the specific soil features observed in the soil profiles, and are detailed in Avery (1980) and in the revised classification of soil series by Clayden and Hollis (1984). In the England and Wales Soil Classification System, soil profile information is defined at four levels in a hierarchy system (Major group–group-subgroup, and series) in descending order, where soil series is more precise in its definition. The national soil map (or

'Natmap') is based on published soil maps which cover a quarter of the land at scales of 1:25,000, 1:63,360 or 1:100,000 and on reconnaissance mapping of previously unsurveyed areas. A total of 67 soil subgroups are recognized. The total area of England (130,395 km^2) and Wales (20,735 km^2) is 151,130 km^2 (62.3% of UK). The total area of the United Kingdom is 242,495 km^2.

1.2.1.5 France Soil Classification System

The French soil classification (CPCS 1967) is derived from the genetic principles inherited from the early Russian pedologists, to which morphological and physico-chemical facts have been added in order to define taxonomic units (Latham 1981). This new system names soil types and links them to a comprehensive reference base (Référentiel pédologique 2008). More than just a soil classification system, it is a coherent method for organizing all the available information. This is considered an effective tool that conveys the necessary information and establishes correlations between different regions.

1.2.1.6 Kuwait Soil Taxonomy Hierarchy

The Soil Survey for the State of Kuwait was completed during 1995–1999 in two phases (KISR 1999a, b; Omar and Shahid 2013). Phase 1 was completed at reconnaissance survey level (4th order level of USDA-NRCS system) at 1:100,000 scale for the entire Kuwait excluding the restricted areas and areas already in use. Phase 2 was completed at semi-detailed (2nd order level of USDA-NRCS system) level at scale 1:25,000 of 200,000 hectares having the potential for irrigated agriculture. A further survey of three demonstration farm sites (50 ha each) was completed (Shahid and Omar 1999; Shahid et al. 2004) at 1:10,000 scale (1st order level of USDA-NRCS system). The USDA-NRCS standards (Soil Survey Division Staff 1993; Soil Survey Staff 1994) were used in soil survey of Kuwait, the results are correlated to the latest standards (Soil Survey Staff 2014; Soil Science Division Staff 2017) and reported in Kuwait Soil Taxonomy. Since the completion of above surveys, no efforts have been made to further soil classification in Kuwait at large or small scales. The publication of Kuwait Soil Taxonomy will be a milestone and fill the gap by providing handy information for potential stakeholders as they strive to better understand and educate the public about this vital natural resource "the soil". Kuwait soil taxonomy hierarchy recognized 2 soil order, 7 suborders, 10 great groups, 18 subgroups, 24 soil families and 39 soil series (KISR 1999a, b; Omar and Shahid 2013).

1.2.1.7 Norway Soil Classification System

In Norway, in 2013 a simplified soil mapping was implemented which was further developed in 2015. The soil classification system of World Reference Base (WRB) for Soil Resources was used during the field work. This soil classification describes a key and list of Norwegian soil series. The Norwegian Institute of Bioeconomy Research (NIBIO) is responsible for international reporting on Norway's soils (https://www.nibio.no/en/subjects/soil/soil-mapping).

1.2.1.8 Russian Soil Classification System

The legend of the Soil Map of the Russian Federated Soviet Republic 1:2.5 M (1988) has been correlated with the Revised FAO and the World Reference Base (WRB) legend. A new version of soil classification in the Soviet Union has been published (Shishov and Sokolov 1992). The knowledge of Russian soils has been intensively reviewed and debated for the past decades (Shishov and Sokolov 1992; Shishov et al. 2001, 2005) and has not yet been completed. The soil map published by Fridland (1988) is considered as standard. Thus, the trend in soil classification development leads to an association of the soil map legend with soil classification in a consistent way. In 2004, the second, refined and complemented version of the national soil classification system—*Classification and Diagnostics of Russian Soils* was published in Russia. The new Russian soil classification system is based on the substantive–genetic principle, i.e., genetically important substantive soil features and properties are taken into account. The second version preserves this principle, as well as the set of criteria used to separate soils at the level of genetic soil type, the central taxon of the system (Shishov et al. 2005). This version was considerably revised and complemented in comparison with the first version published in 1997 (Russian Soil Classification System 1997) and the English translation of this version published in 2001 (Arnold 2001).

1.2.1.9 South African Soil Classification System

The first soil classification system published in 1977 was based on survey information from multiple sources "Soil Classification—a Binomial System for South Africa", known as the "Red book" (Macvicar et al. 1977). In this classification information from top and subsoil horizons was integrated for a specific soil type. The main application of the 1977 system is the national Land Type Survey at 1:250,000 scale. The 1977 soil classification was refined in 1991 (Soil Classification – a Taxonomic System for South Africa" (known as the "blue book") and is commonly used in South Africa (Soil Classification Working Group 1991).

1.2.1.10 United Arab Emirates Keys to Soil Taxonomy

Based on the soil survey work completed in different phases, Dubai emirate (2002–2005, Dubai Municipality 2005), Abu Dhabi emirate (2006–2009, EAD 2009a, b) and Northern emirates of UAE (2010–2012, EAD-MOEW 2012), later Shahid et al. (2014) made significant efforts to publish United Arab Emirates Keys to Soil Taxonomy. The soil classifications in Abu Dhabi emirate soil survey was based on 10th edition (Soil Survey Staff 2006) and Northern emirates of UAE on 11th edition (Soil Survey Staff 2010) of USDA-NRCS Keys to Soil Taxonomy. The field book was used to describe soil horizons of the profile (Schoeneberger et al. 2002, 2012). The United Arab Emirates Keys to Soil Taxonomy (Shahid et al. 2014) is an essential tool to further soil classification studies in the UAE. The development of this valuable piece of work is unique in the Arab region. In this context, the USDA-NRCS recognized this book a basic system that will prove to be very beneficial to anyone wishing to classify and correlate soils within the region.

1.2.1.11 USDA-NRCS Keys to Soil Taxonomy

In the Unites States of America, the Soil Taxonomy was initially developed to support National Cooperative Soil Survey Program. Later, Ditzler and Ahrens (2006) has recognized that the previous system of soil classification was limited in its usefulness and could not be applied consistently by a large cadre of soil scientists with diversified levels of skills and experience. The first edition of Soil Taxonomy was published in 1975 (Soil Survey Staff 1975). Dr. Guy Smith took the leadership to develop this first edition over a period of 25 years. This basic source of soil classification published 45 years ago has been used to support soil survey efforts in many countries including but not necessarily limited to Kingdom of Saudi Arabia (MAW 1985), Sultanate of Oman (MAF 1990), Kuwait (KISR 1999a, b), State of Qatar (Scheibert et al. 2005) and United Arab Emirates (EAD 2009a, b; EAD-MOEW 2012). This basic soil classification system of USA was further revised through number of international committees and many soil scientists, whose work was reflected in 8 editions of Keys to Soil Taxonomy and culminated in the form of the 2nd edition of the full Soil Taxonomy text in the year 1999 (Soil Survey Staff 1999). Later further refinements were made by adding new discoveries of soils and 4 new editions of Keys to Soil Taxonomy were published over a period of 1999 to 2014, the most recently, the 12th edition of the Keys to Soil Taxonomy (Soil Survey Staff 2014).

A recent example of change in USDA-NRCS Soil Taxonomy (Soil Survey Staff 2014) is the recognition of anhydrite ($CaSO_4$) in the soils of United Arab Emirates (Shahid et al. 2007; Wilson et al. 2013) with the addition of a new diagnostic "*Anhydritic*" horizon, a new "*Anhydritic*" mineralogy class, and new "*Anhydritic*" subgroups in the Salids suborder of Aridisols Soil Order, for future use in global soil surveys, where anhydrite may be recognized with the Salic horizon. The USDA-NRCS recognized this addition as significant improvements in soil taxonomy. A continuous change in soil taxonomy since its first publication in 1975 proved it to be a truly universal soil classification system.

1.2.1.12 World Reference Base for Soil Resources (WRB Classification)

Since the nineteenth century, several countries developed national soil classification systems. During the twentieth century, the need for an international soil classification system became more and more obvious. From 1971 to 1981, the Food and Agriculture Organization (FAO) and UNESCO published the Soil Map of the World, 10 volumes, scale 1:5 Million (FAO-UNESCO 1990). The Legend for this map, published in 1974 became the FAO soil classification. The WRB has replaced the FAO/UNESCO Legend for the Soil Map of the World as international standard.

The World Reference Base for Soil Resources (WRB) is an international soil classification system for naming soils and creating legends for soil maps. It was designed to cater for any soil in the world. The WRB is a two-tier system of soil classification, with 32 Major Soil Groups (the "Reference Base") and over 120 uniquely defined qualifiers for specific soil characteristics (the "WRB Classification System"). The currently valid version is the update in 2015 (IUSS Working Group WRB 2015) of the third edition 2014. The WRB borrows heavily from modern soil classification concepts, including Soil Taxonomy, the legend for the FAO Soil Map of the World 1988, the Référentiel Pédologique, and Russian concepts.

References

Abdelfattah MA, Shahid SA (2007) A comparative characterization and classification of soils in Abu Dhabi coastal area in relation to arid and semiarid conditions using USDA and FAO soil classification systems. Arid Land Res Manag 21:245–271

Arnold EW (ed) (2001) Russian soil classification system. Moscow

Avery BW (1980) Soil classification for england and wales (Higher Categories). Soil Survey Technical Monograph No. 14. Harpenden

Baldwin M, Kellogg CE, Thorp J (1938) Soil classification, pp 979–1001. Soils and men: yearbook of agriculture 1938. U.S. Government Printing Office, Washington, D.C

Charles H, Godfray J, Bddington JR, Crute IR, Haddad L, Lawrence D, Muir JF, Pretty J, Robinson S, Thomas SM, Toulmin C (2010) Food security: the challenge of feeding 9 billion peoples. Sciencexpress/www.scienceexpress.org/28January2010/page 1/10.1126/science.1185383

Clayden B, Hollis JM (1984) Criteria for differentiating soil series. Soil Survey Technical Monograph No. 17. Harpenden

CPCS (1967) Classification des sols. Laboratoire des Géologie-Pédologie ENSA Grignon, 87p. Mimeo

Ditzler CA, Ahrens RJ (2006) Development of soil taxonomy in the United States of America. Eurasian Soil Sci 39:141–146

Dubai Municipality (DM) (2005) Satellite imagery and thematic mapping project, 221p. Executed jointly by Global Scan Technologies (Dubai) and National Remote Sensing Agency, India

EAD (2009a) Soil survey of Abu Dhabi Emirate-extensive survey, vol. 1, pp xx+506. Environment Agency–Abu Dhabi, UAE

EAD (2009b) Soil survey of Abu Dhabi Emirate, intensive survey, vol 3, pp xviii+435. Environment Agency, Abu Dhabi

EAD-MOEW (2012) Soil survey of the Northern Emirates. Soil Report, vol 1, pp xxvi+410. Environment Agency–Abu Dhabi, UAE

FAO-UNESCO (1990) Soil map of the world: revised legend. World Soil Report, No. 60. FAO, Rome

Feng LI (2001) Chinese Soil Taxonomy. Science Press Beijing. New York, Coordinated by Institute of Soil Science, Chinese Academy of Sciences, p 203

Fridland VM (ed) (1988) Soil Map of the Russian Soviet Federative Socialist Republic at Scale 1:2.5 Million. All Union Academy of Agricultural Science, Moscow. Government Administration for Geodesy and Cartography (GUGN), 16 sheets

Haynes RH (ed) (1998) The Canadian system of soil classification, 3rd edn. Soil Classification Working Group. Research Branch, Agriculture and Agri-Food Canada. NRC Research Press, Ottawa

Hewitt AE (1992) New Zealand soil classification. DSPIR Land Resources Scientific Report No. 19

Isbell RF (2016) The Australian soil classification, 2nd edn. CSIRO, Collingwood, https://www.clw.csiro.au/aclep/asc_re_on_line_V2/soilbgro.htm

IUSS Working Group WRB (2015) World Reference Base for Soil Resources 2014, update 2015 International soil classification system for naming soils and creating legends for soil maps. World Soil Resources Reports No. 106. FAO, Rome

KISR (1999a) Soil survey for the State of Kuwait: reconnaissance survey, vol II & III. Kuwait Institute for Scientific Research, Kuwait

KISR (1999b) Soil survey for the State of Kuwait: semi-detailed survey, vol IV & V. Kuwait Institute for Scientific Research, Kuwait

Latham M (1981) French soil classification and their application in the South pacific Islands. In: Morrison RJ, Leslie DM (eds) Proceedings of the South Pacific regional forum on soil taxonomy. Institute of Natural Resources, The University of the South Pacific Suva Fiji, pp 185–199

Macvicar CN, De Villiers JM, Loxton RF, Verster E, Lambrechts JNN, Merryweather FR, Le Roux J, Van Rooyen TH, Von HJ, Harmse M (1977) Soil Classification- a binomial system for South Africa. A Report on a Research Project conducted under the Auspices of The Soil and Irrigation Research Institute, Department of Agricultural Technical Services. Republic of South Africa, 150p

MAF (1990) General soil map of the Sultanate of Oman. Ministry of Agriculture and Fisheries, Directorate General of Agricultural Research, Muscat

MAW (1985) General soil map of the Kingdom of Saudi Arabia. Ministry of Agriculture and Water, Riyadh

Moore AW, Isbell RF, Northcote KH (1983) Classification of Australian soils. In Soils: an Australian viewpoint, pp 253–6. (Division of Soils CSIRO). CSIRO, Melbourne/ Academic Press, London

Northcote KH (1979) A factual key for the recognition of Australian soils, 4th edn. Rellim Tech. Publ, Glenside, South Australia

Omar SAS, Shahid SA (2013) Reconnaissance soil survey for the State of Kuwait. In: Shahid SA, Taha FK, Abdelfattah MA (eds) Chapter 3, developments in soil classification, land use planning and policy implications: Innovative thinking of soil inventory for land use planning and management of land resources, pp 85–110

Référentiel pédologique (2008) Afes - Association française pour l'étude des sols (In French)

Russian Soil Classification System (1997) Russian Soil Classification System. Moscow [in Russian]

Scheibert C, Stietiya MH, Sommer J, Abdalla OES, Schramm H, Memah Al M (2005) The atlas of soils for the State of Qatar. Ministry of Municipal Affairs and Agriculture, Doha

Schoeneberger PJ, Wysocki DA, Benham EC, Broderson WD (eds) (2002) Field book for describing and sampling soils, version 2.0. Natural Resources Conservation Service, National Soil Survey Center, Lincoln

Schoeneberger PJ, Wysocki DA, Benham EC, Soil Survey Staff (2012) Field book for describing and sampling soils, Version 3.0. USDA–NRCS, National Soil Survey Center, Lincoln, NE

Shahid SA, Omar SAS (1999) Order 1 soil survey of the demonstration farm sites with proposed management. Kuwait Institute for Scientific Research, Kuwait. viii + 144 pp. KISR 5463. ISBN 0 957700369

Shahid SA, Omar SAS, Jamal ME, Shihab A, Abo-Rezq H (2004) Soil survey for farm planning in northern Kuwait. Kuwait J Sci Engrg 31(1):43–57

Shahid SA, Abdelfattah MA, Wilson M (2007) A unique anhydrite soil in Abu Dhabi. United Arab Emirates. Soil Survey Horizons 48(4):75–79

Shahid SA, Abdelfattah MA, Wilson MA, Kelley JA, Chiaretti JV (2014) United Arab Emirates Keys to Soil Taxonomy, 108p. Springer

Shi X, Yu D, Sun W, Wang H, Zhao Q, Gong Z (2004) Reference benchmarks relating to great groups of genetic soil classification of China with soil taxonomy. Chin Sci Bull 49:1507–1511

Shi XZ, Yu DS, Warner ED, Sun WX, Petersen GW, Gong ZT, Lin H (2006) Cross-reference system for translating between genetic soil classification of China and soil taxonomy. Soil Sci Soc Am J 70(1):78–83

Shi XZ, Yu DS, Xu SX, Warner ED, Wang HJ, Sun WX, Zhao YC, Gong ZT (2010) Cross-reference for relating genetic soil classification of China with WRB at different scales. Geoderma 155(304):344–350

Shishov LL, Sokolov IA (1992) A new version of soil classification in the Soviet Union. Pochvovedenie 4:112–120 [In Russian]

Shishov LL, Tonkonogov VD, Lebedeva II, Gerasimova MI (2001) Principles, structure, and prospects of the new Russian soil classification system. In: Micheli E et al (ed) Soil classification. European Soil Bureau Reports, No. 7. Luxemburg, pp 29–35

Shishov LL, Tonkonogov VD, Lebedeva II, & Gerasimova MI (2004) Classification and diagnostics of Russian soils. Oikumena, Smolensk [in Russian]

Shishov LL, Tonkonogov VD, Gerasimova MI, Lebedeva II (2005) National Soil Classification Systems-New Classification System of Russian Soils. Eurasian Soil Sci 38(Suppl. 1):S35–S43

Soil Classification Working Group (1991) Soil classification a taxonomic system for South Africa. Memoirs on the Agricultural Natural Resources of South Africa No. 15. A Report on a Research Project Conducted under the Auspices of the Soil and Irrigation Research Institute, Department of Agricultural Development, Pretoria, Republic of South Africa, 257p

Soil Classification Working Group (1998) The Canadian system of soil classification, 3rd edn, 187 p. Agriculture and Agri-Food Canada Publication 1646

Soil Survey Division Staff (1993) Soil survey manual. United States Department of Agriculture Handbook No. 18

Soil Science Division Staff (2017) Soil survey manual. United States Department of Agriculture Handbook No. 18

Soil Survey Staff (1975) Soil taxonomy: a basic system of soil classification for making and interpreting soil surveys. Soil Conservation Service. US Department of Agriculture Handbook 436

Soil Survey Staff (1999) Soil taxonomy: a basic system of soil classification for making and interpreting soil surveys, 2nd edn. Natural Resources Conservation Service. US Department of Agriculture Handbook 436

Soil Survey Staff (1994) Keys to soil taxonomy, 6th edn. USDA-NRCS

Soil Survey Staff (2006) Keys to soil taxonomy, 10th edn. USDA-NRCS

Soil Survey Staff (2010) Keys to soil taxonomy, 11th edn. USDA-NRCS

Soil Survey Staff (2014) Keys to soil taxonomy, 12th edn. USDA-NRCS

Stace HCT, Hubble GD, Brewer R, Northcote KH, Sleeman JR, Mulcahy MJ, Hallsworth EG (1968) A handbook of Australian soils. Rellim Tech. Pubs., Glenside, S.A

Wilson MA, Shahid SA, Abdelfattah MA, Kelley JA, Thomas JE (2013) Anhydrite formation on the coastal sabkha of Abu Dhabi, United Arab Emirates. In: Shahid SA, Taha FK, Abdelfattah MA (eds) Chapter 8: developments in soil classification, land use planning and policy implications-innovative thinking of soil inventory for land use planning and management of land resources. Springer, pp 175–202

Seidlitzia rosmarinus is growing in saline soil at low elevation

2 The Soil That We Attempt To Classify

Abstract

The soil is a mixture of mineral and organic particles of various sizes, water and the air. The proportion of these soil components determine the physical makeup and other properties of the soil. The mineral matter constitutes of three primary soil particles based on their equivalent spherical diameter (esd), sand (2–0.05 mm), silt (0.05–0.002 mm) and the clay (<0.002 mm). The percent distribution of these three primary particles determines the soil texture class. There are 12 soil textural classes in the USDA system. In Kuwait major soil texture classes are sand and loamy-sand. At the global and national levels, there is wide diversity of soils with different physical, chemical, mineralogical and biological properties. This diversity makes soils one of the most complex natural systems that what we attempt to classify for informed decisions for land use planning and national development. The lower boundary of soil that we classify is arbitrary set at 200 cm. Soil, as defined in this text, does not need to have discernible genetic horizons, although the presence or absence of genetic horizons and their nature are of extreme importance in soil classification. The Kuwait Soil Taxonomy is presented in a way to key out soils starting from the beginning and eliminating one by one not matching the criteria to the soil in

S. A. Shahid and S. A. S. Omar, *Kuwait Soil Taxonomy*,
https://doi.org/10.1007/978-3-030-95297-6_2

the field until correct name of surface and subsurface horizons are identified and a correct soil taxa name is designated to the location under investigation.

Keywords
Buried soil · Soil formation · Lower boundary · Classification · Taxonomy

2.1 Introduction

Soils are one of the most important components for agrifood systems, but also one of the most fragile: the formation of a centimeter of land requires hundreds of years of work, and can be destroyed in seconds. Therefore, we must encourage conservation practices and the adoption of technologies to restore degraded land and protect those that are still functional. This requires proper understanding of national soils using internationally recognized soil classification system which fits the countries climatic conditions and covers the soil formation processes and soil taxonomic classes.

The word "soil" has diversified meanings based on its use for a specific purpose. For a potter who makes and sell soil based pottery, it is a source material for pots production, for a civil engineer, soil is construction material (houses, road beds etc.) and foundations for buildings, soil also provides materials for ceramics, and medium for industrial and waste disposal, protects groundwater by filtering toxic chemicals etc.. From the point of view of an agriculturist, it is "natural medium for the growth of land plants, whether or not it has discernible soil horizons or layers". The soil supports plants that supply food, fibers, drugs, and other wants of humans and because it filters water and recycles wastes. In addition, the soils are considered dynamic ecosystems capable of supporting plant and wildlife. Soil is considered "skin" of the planet earth. Soil formation (Jenny 1941) is a function of climate (cl), organism (o), relief (r), parent material (p) and time (t) and expressed as $S = f\,(cl, o, r, p, t)$.

National soil classification is essential to understand the history of soil formation, explaining the distribution and heterogeneity on the landscape, understanding soil properties, soil formation processes and modeling, land use evaluation and relationship with the environment and for the provision of ecosystem services.

2.2 The Soil That We Classify

Soil in this text is a natural body comprised of solids (minerals and organic matter), liquid, and gases that occur on the land surface, occupies space, and is characterized by one or both of the following: horizons, or layers, that are distinguishable from the initial material as a result of additions, losses, transfers, and transformations of energy and matter *or* the ability to support rooted plants in a natural environment (Soil Survey Staff 1999). This definition is expanded from the previous version of *Soil Taxonomy* (Soil Survey Staff 1975). The upper limit of soil is the boundary between soil and either air, shallow water, live plants, or plant materials that have not begun to decompose. Areas are not considered to have soil if the surface is permanently covered by water too deep (typically more than about 2.5 m) for the growth of rooted plants. The horizontal boundaries of soil are areas where the soil grades to deep water, barren areas, rock, or ice. Soil is not static, but dynamic. The pH, soluble salts, amount of organic matter and carbon–nitrogen ratio, numbers of microorganisms, soil fauna, temperature, and moisture status all change with the seasons as well as with more extended periods of time. Soil must be viewed from both the short-term and long-term perspective (Soil Survey Staff 2014).

2.3 Lower Boundary of Soil

Soil consists of the horizons near the earth's surface that, in contrast to the underlying parent material, have been altered by the interactions of climate, relief, and living organisms over time (Jenny 1941). For the practicality of soil survey, the lower boundary of soil is arbitrarily set at 200 cm. In areas where soil has thin, pedogenically cemented horizons that are impermeable to roots, the soil extends as deep as the deepest cemented horizon, but not below 200 cm. For certain management goals, layers deeper than the lower boundary of the soil that is classified (200 cm) must also be described if they affect the content and movement of water and air or other interpretative concerns.

2.4 Nonsoil Materials

Plants can be grown under glass in pots filled with earthy materials, such as peat or sand, or even in water. Under proper conditions all these media are productive for plants, but they are nonsoil here in the sense that they cannot be classified in the same system that is used for the soils of a survey area, Governorate, or even nation. Plants even grow on trees or in cracks of exposed bedrock (i.e., rock outcrop), but trees and rock outcrop are regarded as nonsoil.

2.5 Buried Soils

A buried soil is a sequence of genetic horizons in a pedon that is covered with a surface mantle of new soil material, that is 50 cm or more thick. In Kuwait, the mobile sand usually covers the genetic horizons, thus can be considered a buried soil.

The hierarchy of soil classification (KISR 1999a, b) has been used to prepare Kuwait Soil Taxonomy. It is hoped that this key will be widely used by the potential soil information seekers to find locations for specific purposes.

References

Jenny H (1941) Factors of soil formation. McGraw-Hill, New York, London

KISR (1999a) Soil survey for the state of Kuwait: reconnaissance survey, vol II & III. Kuwait Institute for Scientific Research, Kuwait

KISR (1999b) Soil survey for the state of Kuwait: semi-detailed survey, vol IV & V. Kuwait Institute for Scientific Research, Kuwait

Soil Survey Staff (1975) Soil taxonomy: a basic system of soil classification for making and interpreting soil surveys. Soil Conservation Service. U.S. Department of Agriculture Handbook, vol 436

Soil Survey Staff (1999) Soil taxonomy: a basic system of soil classification for making and interpreting soil surveys, 2nd edn. Natural Resources Conservation Service. U.S. Department of Agriculture Handbook, vol 436

Soil Survey Staff (2014) Keys to soil taxonomy, 12th edn. US Department of Agriculture, Natural Resources Conservation Service, US Government Printing Office, Washington, DC

Halocnemum strobilaceum grows in the littoral salt marshes inundated by sea water

3 Horizons, Layers, and Characteristics Diagnostics for the Higher Categories of Soil Classification in Kuwait

Abstract

In this chapter we defined the master and diagnostic horizons identified in the soils of Kuwait. Among the master horizons, in Kuwait A, B, C and R have been identified in both the reconnaissance and semi-detailed survey of Kuwait. Whereas, eight diagnostic horizons (argillic, calcic, cambic, gypsic, ochric, petrocalcic, petrogypsic and salic) have been identified. Each master horizon or soil layer is allocated one or more small letters called suffix, which has specific meaning to describe the characteristics of the master horizon. In Kuwait, b, k, m, t, y and z are used. The Aquic, Aridic or Torric moisture regime and hyperthermic temperature regime are identified in Kuwait. The horizons and characteristics defined in this chapter are not in a key format.

Keywords

Horizons · Suffixes · Discontinuities · Epipedon · Diagnostic horizons

3.1 Introduction

In this chapter we defined the characteristics of the genetic, master horizons and layers identified and reported in the soil survey of Kuwait reports (KISR 1999a, b; Omar and Shahid 2013). The information relevant to the soils of Kuwait (Aridisols

S. A. Shahid and S. A. S. Omar, *Kuwait Soil Taxonomy*,
https://doi.org/10.1007/978-3-030-95297-6_3

and Entisols soil orders, see Chap. 5 for more details) are extracted from the latest Keys to Soil Taxonomy (Soil Survey Staff 2014) with slight changes to fit soil classification of Kuwait. An official permission has been granted by the USDA management to extract information relevant to the soils of Kuwait.

3.2 Basic System of Horizon and Layer Designations

While describing the morphological features of the soil profiles and/or auger holes to classify a soil, four kinds of symbols are used in various combinations to designate properties of horizons and layers:

- Capital letters (A, E, B, C, R etc.) are allocated to the master horizons and layers.
- Lowercase letters (small letters) are used as suffixes (e.g., Ak) to describe specific characteristics of master horizons and layers.
- Numbers are used both as suffixes to indicate vertical subdivisions (e.g., Ak1, Ak2, Ak3 etc.) within a horizon or layer and as prefixes to indicate discontinuities.
- Special symbols are used to indicate layers formed in human-transported material or sequences of horizons having otherwise identical designations.

3.2.1 Master Horizons and Layers

In the USDA-NRCS (Soil Survey Staff 2014; Soil Science Division Staff 2017) system of soil classification, the capital letters A, B, C, R represent the master horizons and layers of soils identified in Kuwait. Each horizon-layer has specific definition. These letters are the base symbols to which other characters are added to complete the designations. Most horizons and layers are given a single capital-letter symbol; some require two. In this section, preference is given to the horizons and layers identified in Kuwait, these are: A, B, C and R. The definitions are restricted to the features observed in Kuwait, rather than giving detailed explanation to represent properties in other soils not related to Kuwait.

A Horizons

A horizons are mineral horizons that formed at the soil surface or below an O (organic) horizon. It may contain humified organic matter mixed with mineral matter and show signs of cultivation.

B horizons

Mineral horizons formed below an A, or E, it shows signs of illuvial accumulation of clay, carbonates, gypsum etc.) demonstrating the operation of soil genesis (pedogenesis) process and have structure other than rock structure.

C horizon

C horizons or layers lack the properties of O, A, E and B horizons and are little affected by pedogenesis.

R Layers

R layers consist of strongly cemented to indurated bedrock. Limestone and sandstone are examples of bedrock designated by the letter R.

In the south of Kuwait deep sandy soils are identified. These soils do not show evidence of pedogenesis, such as accumulation of clay, carbonates and gypsum. However, in the north of Kuwait the pedogenesis has been observed in the form of distinct accumulation of secondary carbonates and gypsum. In the west of Kuwait gypsic hardpan soils have been mapped. The lithified layers (sandstone) are designated as R.

3.2.2 Suffix Symbols

The master horizons and layers (A, B, C, R) also show lowercase letters (such as Ak, By) called suffix symbols. These symbols are used to designate specific subordinate distinctions within master horizons and layers. In Kuwait, b, k, m, t, y and z are used. Other suffix symbols (kk, n, p, w, and yy) were not recorded, but are likely to occur in Kuwait and may be useful to record for future work. These are described below:

b *Buried genetic horizon*

This symbol indicates identifiable buried horizons with major genetic features that were developed before burial. Genetic horizons may or may not have formed in the overlying material, which may be either like or unlike the assumed parent material of the buried horizon.

In Kuwait mobile sand in the desert, in some places has buried the genetic horizon, such a feature was observed in the Wafra area, where calcic genetic horizon was buried by a thick layer of sandy surface mantle.

k *Accumulation of secondary carbonates*

This symbol indicates an accumulation of visible pedogenic calcium carbonate (less than 50 percent, by volume). Carbonate accumulations in the soils of Kuwait were found in the form of disseminated carbonates and secondary calcium carbonates accumulations. The symbol k can be used with horizons showing very slightly to strong effervescence with dilute hydrochloric acid.

kk *Engulfment of horizon by secondary carbonates*

This symbol indicates major accumulations of pedogenic calcium carbonate. The suffix kk is used when the soil fabric is plugged with fine grained pedogenic carbonate (50 percent or more, by volume) that occurs as an essentially continuous medium. The suffix corresponds to stage III of the carbonate morphogenetic stages (Gile et al. 1966) or a higher stage.

m *Pedogenic cementation*

This symbol indicates continuous or nearly continuous pedogenic cementation. It is used only for horizons that are more than 90 percent cemented but may be fractured. The cemented layer is physically root-restrictive. The predominant cementing agent (or the two dominant ones) can be indicated by letter suffixes, singly or in pairs. The horizon suffix km indicates cementation by carbonates and ym indicates cementation by gypsum.

n *Accumulation of sodium*

This symbol indicates an accumulation of exchangeable sodium.

p *Tillage or other disturbance*

This symbol indicates a disturbance of a horizon by mechanical means, ploughing, or similar uses. A disturbed mineral horizon is designated Ap even though it is clearly a former B or C horizon.

r *Weathered or soft bedrock.*

This symbol is used with C horizons to indicate layers of bedrock that are moderately cemented or less cemented. Examples are weathered igneous rock and partly consolidated sandstone, siltstone, or shale. The excavation difficulty is low to high.

t *Accumulation of silicate clay*

This symbol indicates an accumulation of silicate clay that either has formed within a horizon and subsequently has been translocated within the horizon or that has been moved into the horizon by illuviation, or both. At least some part of the horizon shows evidence of clay accumulation, either as coatings on surfaces of peds or in pores, as lamellae, or as bridges between mineral grains.

w *Development of color or structure*

This symbol is used only with B horizons to indicate the development of color or structure, or both, with little or no apparent illuvial accumulation of material. It should not be used to indicate a transitional horizon.

y *Accumulation of gypsum*

This symbol indicates an accumulation of gypsum. It is used when the horizon fabric is dominated by soil particles or minerals other than gypsum. Gypsum is present in amounts that do not significantly obscure or disrupt other features of the horizon. This symbol is also used to indicate the presence of anhydrite.

yy *Dominance of horizon by gypsum*

This symbol indicates a horizon that is dominated by the presence of gypsum. The gypsum content may be due to an accumulation of secondary gypsum, the transformation of primary gypsum inherited from parent material, or other processes. The suffix yy is used when the horizon fabric has such an abundance of gypsum (generally 50 percent or more, by volume) that pedogenic and/or lithologic features are obscured or disrupted by growth of gypsum crystals. Colors associated with horizons that have suffix yy typically are highly whitened (e.g., value of 7 through 9.5 and chroma of 4 or less). This symbol is also used to connote the presence of anhydrite ($CaSO_4$).

z *Accumulation of salts more soluble than gypsum*

This symbol indicates an accumulation of salts that are more soluble than gypsum ($CaSO_4.2H_2O$) and anhydrite ($CaSO_4$). It is commonly used when ECe is more than 4 dS/m.

3.2.3 Conventions for Using Horizon Designation Symbols

Many master horizons and layers that are symbolized by a single capital letter can have one or more lowercase-letter suffixes. The following guides can be used.

1. Letter suffixes directly follow the capital letter of the master horizon or layer (eg., Ak. By)
2. More than three suffixes are rarely used.
3. If more than one suffix is needed, the following letters (if used) are written first: *a, d, e, h, i, r, s, t,* and *w*. None of these letters are used in combination for a single horizon.
4. If more than one suffix is needed and the horizon is not buried, the following symbols, if used, are written last: *c, f, g, m, v,* and *x*. Examples from Kuwait is Bkym. If any of these suffixes are used together in the same horizon, symbols *c* and *g* are written last (e.g., Btg).
5. If a genetic horizon is buried, the suffix *b* is written last, e.g., Btb

A B horizon that has a significant accumulation of clay and also shows development of color or structure, or both, is designated “Bt” (suffix symbol *t* has precedence over symbols *w, s,* and *h*). A B horizon that is gleyed or that has accumulations of carbonates, sodium, silica, gypsum, salts more soluble than gypsum, or residual accumulations of sesquioxides carries the appropriate symbol: *g, k, kk, n, q, y, yy, z,* or *o*. If illuvial clay is also present, the symbol *t* precedes the other symbol, e.g., Btk, Bty.

3.2.4 Vertical Subdivisions

Commonly, a horizon or layer designated by a single letter or a combination of letters has to be subdivided. For this purpose, numbers are added to the letters of the horizon designation. These numbers follow all the letters. Within a sequence of C horizons, for example, successive layers may be designated C1, C2, C3, etc.

These conventions apply regardless of the purpose of the subdivision. In many soils a horizon that could be identified by a single set of letters is subdivided to recognize differences in morphological features, such as structure, color, or texture. These divisions are numbered consecutively, but the numbering starts again at 1 when any letter of the horizon symbol changes, e.g., Bk1-Bk2-Bky1-Bky2 (not Bk1-Bk2-Bky3-Bky4), Bk1-Bk2-Bk3-Bkm1-Bkm2 (not Bk1-Bk2-Bk3-Bkm4-Bkm5). The numbering of vertical subdivisions within consecutive horizons is not interrupted at a discontinuity (indicated by a numerical prefix) if the same letter combination is used in both materials, e.g. By1-By2-2By3 (not By1-By2-2By1) and Bk1-Bk2-Bk3-2BK4-2BK5 (not Bk1-Bk2-Bk3-2Bk1-2Bk2).

During sampling for laboratory analyses, thick soil horizons are sometimes subdivided even though differences in morphology are not evident in the field.

These subdivisions are identified by numbers that follow the respective horizon designations. For example, four subdivisions of a Bk horizon sampled by 10-cm increments are designated Bk1, Bk2, Bk3, and Bk4. If the horizon has already been subdivided because of differences in morphological features, the set of numbers that identifies the additional sampling subdivisions follows the first number. For example, three subdivisions of a Bk2 horizon sampled by 10-cm increments are designated Bk21, Bk22, and Bk23. The descriptions for each of these sampling subdivisions can be the same, and a statement indicating that the horizon has been subdivided only for sampling purposes can be added.

3.2.5 Discontinuities

Numbers are used as prefixes to horizon designations (specifically, A, E, B, C, and R) to indicate discontinuities in mineral soils. These prefixes are distinct from the numbers that are used as suffixes denoting vertical subdivisions.

A discontinuity that can be identified by a number prefix is a significant change in particle-size distribution or mineralogy that indicates a difference in the parent material from which the horizons have formed and/or a significant difference in age, unless the difference in age is indicated by the suffix *b*. Symbols that identify discontinuities are used only when they can contribute substantially to an understanding of the relationships among horizons. The stratification common to soils that formed in alluvium is not designated as a discontinuity, unless particle-size distribution differs markedly from layer to layer (i.e., particle-size classes are strongly contrasting) even though genetic horizons may have formed in the contrasting layers.

If a soil formed entirely in one kind of material, the whole profile is understood to be material 1 and the number prefix is omitted from the symbol. Similarly, the uppermost material in a profile consisting of two or more contrasting materials is understood to be material 1 and the number is omitted. Numbering starts with the second layer of contrasting material, which is designated 2. Underlying contrasting layers are numbered consecutively. Even when the material of a layer below material 2 is similar to material 1, it is designated 3 in the sequence; the numbers indicate a change in materials, not types of material. Where two or more consecutive horizons have formed in the same kind of material, the same prefix number indicating the discontinuity is applied to all the designations of horizons in that material, for example, Ak-Bk1-Bk2-Bk3-2Bk4-2Bk5. The suffix numbers designating vertical subdivisions of the Bt horizon continue in consecutive order across the discontinuity. However, vertical subdivisions do not continue across lithologic discontinuities if the horizons are not consecutive or contiguous to each other. If other horizons intervene, another vertical numbering sequence begins for the lower horizons, for example, Ak-Bk1-Bk2-Bk3-2Cy-2C.

If an R layer is below a soil that formed in residuum and if it is similar to the material from which the soil developed, the number prefix is not used. The prefix is used, however, if it is thought that the R layer would weather to material unlike that in the solum, e.g., Ak-Bk-C-2R, Ak-Bk-BC-2R, Ak-Cr-2R.

A buried genetic horizon (designated by the suffix *b*) requires special consideration. It is obviously not in the same deposit as the overlying horizons. Some buried horizons, however, formed in material that is lithologically like the overlying deposit. In this case, a prefix is not used to distinguish material of the buried horizon. If the material in which a horizon of a buried soil formed is lithologically unlike the overlying material, the discontinuity is indicated by a number prefix and the symbol for the buried horizon also is used, for example, Ak-Bk1-Bk2-C-2Btb.

3.2.6 The Prime Symbol

If two or more horizons with identical number prefixes and letter combinations are separated by one or more horizons with a different horizon designation, identical letter and number symbols can be used for those horizons with the same characteristics. For example, the sequence A-E-Bt-E-Btx-C identifies a soil that has two E horizons. To emphasize this characteristic, the prime symbol (′) is added after the symbol of the lower of the two horizons that have identical designations, e.g., A-E-Bt-E′-Btx-C. The prime symbol is placed after the master horizon symbol and before the suffix letter symbol or symbols (if used), for example, B′t.

The prime symbol is not used unless all letters and number prefixes are completely identical. The sequence A-Bt1-Bt2-2E-2Bt1-2Bt2 is an example. Because it has two Bt master horizons of different lithologies, the Bt horizons are not identical and the prime symbol is not needed. The prime symbol is used for soils with lithologic discontinuities if horizons have identical designations. In the rare cases where three layers have identical letter symbols, double prime symbols can be used for the lowest of these horizons, for example, E″.

3.2.7 The Caret Symbol

The caret symbol (^) is used as a prefix to indicate horizons and layers that formed in human-transported material. This material has been moved horizontally onto a pedon from a source area outside of that pedon by purposeful human activity, usually with the aid of machinery or hand tools. Number prefixes may be used before the caret symbol to indicate the presence of discontinuities within the human-transported material (e.g., ^Au-^Bwu-^BCu-2^Cu1-2^Cu2) or between the human-transported material and underlying horizons formed in other parent materials (e.g., ^A-^C1-2^C2-3Bwb).

We have given the above explanation about the Prime and Caret symbols, in case the surveyors in Kuwait come across such features in future soil classification studies. But, in the previous surveys (KISR 1999a, b), these symbols are not used.

3.3 Diagnostic Surface and Subsurface Horizons

3.3.1 The Epipedon

The epipedon (Gr. *epi*, over, upon, and *pedon*, soil) is a horizon that forms at or near the surface and in which most of the rock structure has been destroyed. It is darkened by organic matter or shows evidence of eluviation, or both. *Rock structure* as used here and in other places in this taxonomy includes fine stratification (5 mm or less thick) in unconsolidated sediments (aeolian, alluvial, lacustrine, or marine) and saprolite derived from consolidated rocks in which the unweathered minerals and pseudomorphs of weathered minerals retain their relative positions to each other.

Any horizon may be at the surface of a truncated soil. These horizons can be covered by a surface mantle of new soil material. If the surface mantle has rock structure, the top of the epipedon is considered the soil surface unless the mantle meets the definition of buried soils. If the soil includes a buried soil, the epipedon, if any, is at the soil surface and the epipedon of the buried soil is considered a buried epipedon and is not considered in selecting taxa unless the keys specifically indicate buried horizons. A soil with a mantle thick enough to have a buried soil has no epipedon if the soil has rock structure to the surface or has an Ap horizon less than 25 cm thick that is underlain by soil material with rock structure.

A recent alluvial or aeolian deposit that retains fine stratifications (5 mm or less thick) or an Ap horizon directly underlain by such stratified material is not included in the concept of the epipedon because time has not been sufficient for soil-forming processes to erase these transient marks of deposition and for diagnostic and accessory properties to develop. An epipedon is not the same as an A horizon. It may include part or all of an illuvial B horizon if the darkening by organic matter extends from the soil surface into or through the B horizon.

3.3.1.1 Ochric Epipedon

The US Keys to Soil Taxonomy (Soil Survey Staff 2014) has reported eight epipedons (Anthropic, folistic, histic, melanic, mollic, ochric, plaggen, and umbric). In Kuwait only one "Ochric" surface horizon was identified, which is given below.

The ochric epipedon fails to meet the definitions for any of the other seven epipedons because it is too thin or too dry, has too high a color value or chroma, contains too little organic carbon, has too high an *n* value or melanic index, or is both massive and hard or harder when dry. The ochric epipedon includes eluvial horizons that are at or near the soil surface, and it extends to the first underlying diagnostic illuvial horizon. The ochric epipedon does not have rock structure and does not include finely stratified fresh sediments, nor can it be an Ap horizon directly overlying such deposits.

3.3.2 Diagnostic Subsurface Horizons

The horizons described in this section form below the surface of the soil. They are composed of mineral soil material. They may be exposed at the surface by truncation of the soil. Some of these horizons are designated as B horizons by many, but not all pedologists, and others are generally designated as parts of A or E horizons.

Among the 20 subsurface diagnostic horizons (Agric, albic, anhydritic, argillic, calcic, cambic, duripan, fragipan, glossic, gypsic, kandic, natric, orstein, oxic, petrocalcic, petrogypsic, placic, salic, sombric and spodic) reported by the US Keys to Soil Taxonomy (Soil Survey Staff 2014), in Kuwait only seven (Argillic, calcic, cambic, gypsic, petrocalcic, petrogypsic, salic) have been identified, however anhydritic horizon is also added as it is likely to be identified in future soil surveys in Kuwait.

3.3.2.1 Anhydritic Horizon

The anhydritic diagnostic horizon is introduced in the US Keys to soil taxonomy in 2014 (Soil Survey Staff 2014). It was first recognized in United Arab Emirates (Shahid et al. 2007, 2014; Wilson et al. 2013). It was not reported in the soil survey of Kuwait, but likely to be identified in Kuwait due to a large area occupied by gypsum accumulation. Anhydrite ($CaSO_4$) is a dehydrated form of gypsum ($CaSO_4.2H_2O$), there is high possibility of its occurrence in Kuwait, therefore, it is added in the Kuwait Soil Taxonomy to aid soil surveyors in future correct separation of anhydritic horizon from gypsic horizon. It is assumed that anhydritic horizon may have been identified but due to its absence in US Keys to soil taxonomy, it was classified under more close gypsum category.

The anhydritic horizon is a horizon in which anhydrite has accumulated through neoformation or transformation to a significant extent. It typically occurs as a subsurface horizon. It commonly occurs in conjunction with a salic horizon (defined below).

Required Characteristics

The anhydritic horizon meets *all* of the following requirements:

1. Is 15 cm or more thick; *and*
2. Is 5 percent or more (by weight) anhydrite; *and*
3. Has hue of 5Y, chroma (moist and dry) of 1 or 2, and value of 7 or 8; *and*
4. Has a product of thickness, in cm, multiplied by the anhydrite content (percent by weight) of 150 or more (thus, a horizon 30 cm thick that is 5 percent anhydrite qualifies as an anhydritic horizon); *and*
5. Has anhydrite as the predominant calcium sulfate mineral with gypsum either absent or present only in minor amounts.

3.3.2.2 Argillic Horizon

An argillic horizon is normally a subsurface horizon with a significantly higher percentage of phyllosilicate clay than the overlying soil material. It shows evidence of clay illuviation. The argillic horizon forms below the soil surface, but it may be exposed at the surface later by erosion.

Required Characteristics

1. All argillic horizons must meet *both* of the following requirements:

   a. *One* of the following:

      (1) If the argillic horizon meets the particle-size class criteria for coarse-loamy, fine-loamy, coarse-silty, fine-silty, fine, or very-fine or is loamy or clayey, including skeletal counterparts, it must be at least 7.5 cm thick or at least one-tenth as thick as the sum of the thickness of all overlying horizons, whichever is greater; *or*

      (2) If the argillic horizon meets the sandy or sandy-skeletal particle-size criteria, it must be at least 15 cm thick; *or*

(3) If the argillic horizon is composed entirely of lamellae, the combined thickness of the lamellae that are 0.5 cm or more thick must be 15 cm or more; *and*

b. Evidence of clay illuviation in at least *one* of the following forms:

(1) Oriented clay bridging the sand grains; *or*
(2) Clay films lining pores; *or*
(3) Clay films on both vertical and horizontal surfaces of peds; *or*
(4) Thin sections with oriented clay bodies that are more than 1 percent of the section; *or.*
(5) If the coefficient of linear extensibility is 0.04 or higher and the soil has distinct wet and dry seasons, then the ratio of fine clay to total clay in the illuvial horizon is greater by 1.2 times or more than the ratio in the eluvial horizon; *and*

2. If an eluvial horizon remains and there is no lithologic discontinuity between it and the illuvial horizon and no plow layer directly above the illuvial layer, then the illuvial horizon must contain more total clay than the eluvial horizon within a vertical distance of 30 cm or less, as follows:

a. If any part of the eluvial horizon has less than 15 percent total clay in the fine-earth fraction (<2 mm), the argillic horizon must contain at least 3 percent (absolute) more clay (10 percent versus 13 percent, for example); *or*
b. If the eluvial horizon has 15 to 40 percent total clay in the fine-earth fraction, the argillic horizon must have at least 1.2 times more clay than the eluvial horizon; *or*
c. If the eluvial horizon has 40 percent or more total clay in the fine-earth fraction, the argillic horizon must contain at least 8 percent (absolute) more clay (42 percent versus 50 percent).

3.3.2.3 Calcic Horizon

The calcic horizon (Fig. 3.1) is an illuvial horizon in which secondary calcium carbonate or other carbonates have accumulated to a significant extent.

Fig. 3.1 Calcic diagnostic horizon in a *Typic Haplocalcids* profile. Calcium carbonates precipitation in soil matrix is evident

Required Characteristics

The calcic horizon:

1. Is 15 cm or more thick; *and*
2. Has *one or more* of the following:

   a. 15 percent or more (by weight, fine-earth fraction) $CaCO_3$ equivalent, and its $CaCO_3$ equivalent is 5 percent or more (absolute) higher than that of an underlying horizon; *or*
   b. 15 percent or more (by weight, fine-earth fraction) $CaCO_3$ equivalent and 5 percent or more (by volume) identifiable secondary carbonates; *or*
   c. 5 percent or more (by weight, fine-earth fraction) calcium carbonate equivalent and:

      (1) Has less than 18 percent clay in the fine-earth fraction; *and*
      (2) Meets the criteria for a sandy, sandy-skeletal, coarse-loamy, or loamy-skeletal particle-size class; *and*

(3) Has 5 percent or more (by volume) identifiable secondary carbonates or a calcium carbonate equivalent (by weight, fine-earth fraction) that is 5 percent or more (absolute) higher than that of an underlying horizon; *and*

3. Is not cemented or indurated in any part by carbonates, with or without other cementing agents, or is cemented in some part and the cemented part satisfies *one* of the following:

   a. It is characterized by so much lateral discontinuity that roots can penetrate through noncemented zones or along vertical fractures with a horizontal spacing of less than 10 cm; *or*
   b. The cemented layer is less than 1 cm thick and consists of a laminar cap underlain by a lithic or paralithic contact; *or*
   c. The cemented layer is less than 10 cm thick.

3.3.2.4 Cambic Horizon

A cambic horizon is the result of physical alterations, chemical transformations, or removals or of a combination of two or more of these processes.

Required Characteristics

The cambic horizon is an altered horizon 15 cm or more thick. If it is composed of lamellae, the combined thickness of the lamellae must be 15 cm or more. In addition, the cambic horizon must meet *all* of the following:

1. Has a texture class of very fine sand, loamy very fine sand, or finer; *and.*
2. Shows evidence of alteration in *one* of the following forms:

   a. Aquic conditions within 50 cm of the soil surface or artificial drainage and *all* of the following:

      (1) Soil structure or the absence of rock structure, including fine stratifications (5 mm or less thick), in more than one-half of the volume; *and*
      (2) Colors that do not change on exposure to air; *and.*
      (3) Dominant color, moist, on faces of peds or in the matrix as follows:

(a) Value of 3 or less and neutral colors with no hue (N) and zero chroma; *or*
(b) Value of 4 or more and chroma of 1 or less; *or*
(c) Any value, chroma of 2 or less, and redox concentrations; *or*

b. Does not have the combination of aquic conditions within 50 cm of the soil surface or artificial drainage and colors, moist, as defined in item 2-a-(3) above, and has soil structure or the absence of rock structure, including fine stratifications (5 mm or less thick), in more than one-half of the volume and *one or more* of the following properties:

(1) Higher chroma, higher value, redder hue, or higher clay content than the underlying horizon or an overlying horizon; *or*
(2) Evidence of the removal of carbonates or gypsum; *and.*

3. Has properties that do not meet the requirements for an anthropic, histic, folistic, melanic, mollic, plaggen, or umbric epipedon, a duripan or fragipan, or an argillic, calcic, gypsic, natric, oxic, petrocalcic, petrogypsic, placic, salic, spodic, or sulfuric horizon; *and*
4. Is not part of an Ap horizon and does not have a brittle manner of failure in more than 60 percent of the matrix.

3.3.2.5 Gypsic Horizon

The gypsic horizon is a horizon in which gypsum has accumulated or been transformed to a significant extent. It typically occurs as a subsurface horizon, but it may occur at the surface in some soils.

Required Characteristics

A gypsic horizon (Fig. 3.2) meets *all* of the following requirements:

1. Is 15 cm or more thick; *and*
2. Is not cemented by gypsum, with or without other cementing agents; is cemented and the cemented parts are less than 5 mm thick; or is cemented but, because of lateral discontinuity, roots can penetrate along vertical fractures with a horizontal spacing of less than 10 cm; *and*
3. Is 5 percent or more (by weight) gypsum and has 1 percent or more (by volume) visible secondary gypsum that has either accumulated or been transformed; *and*

Fig. 3.2 Gypsic diagnostic horizon in a *Leptic Haplogypsids* profile. Gypsum accumulation starting at 15 cm is evident

4. Has a product of thickness, in cm, multiplied by the gypsum content (percent by weight) of 150 or more. Thus, a horizon 30 cm thick that is 5 percent gypsum qualifies as a gypsic horizon if it is 1 percent or more (by volume) visible gypsum and any cementation is as described in 2 above.

3.3.2.6 Petrocalcic Horizon

The petrocalcic horizon is an illuvial horizon in which secondary calcium carbonate or other carbonates have accumulated to the extent that the horizon is cemented or indurated.

Required Characteristics

A petrocalcic horizon must meet the following requirements:

1. The horizon is cemented or indurated by carbonates, with or without silica or other cementing agents; *and*
2. Because of lateral continuity, roots can penetrate only along vertical fractures with a horizontal spacing of 10 cm or more; *and*
3. The horizon has a thickness of:

a. 10 cm or more; *or*
b. 1 cm or more if it consists of a laminar cap directly underlain by bedrock.

3.3.2.7 Petrogypsic Horizon

The petrogypsic horizon (Fig. 3.3) is a horizon in which visible secondary gypsum has accumulated or has been transformed. The horizon is cemented (i.e., extremely weakly cemented through indurated cementation classes), and the cementation is both laterally continuous and root limiting, even when the soil is moist. The horizon typically occurs as a subsurface horizon, but it may occur at the surface in some soils.

Required Characteristics

A petrogypsic horizon meets *all* of the following requirements:

1. Is cemented or indurated by gypsum, with or without other cementing agents; *and*
2. Because of lateral continuity, can be penetrated by roots only along vertical fractures with a horizontal spacing of 10 cm or more; *and*
3. Is 5 mm or more thick; *and*
4. Is 40 percent or more (by weight) gypsum.

3.3.2.8 Salic Horizon

A salic horizon is a horizon of accumulation of salts that are more soluble than gypsum in cold water (Fig. 3.4).

Required Characteristics

A salic horizon is 15 cm or more thick and has, for 90 consecutive days or more in normal years:

1. An electrical conductivity (EC) equal to or greater than 30 dS/m in the water extracted from a saturated paste; *and*
2. A product of the EC, in dS/m, and thickness, in cm, equal to 900 or more.

Fig. 3.3 Petrogypsic diagnostic horizon in a *Typic Petrogypsids* profile starting at 60 cm depth

Fig. 3.4 A salic diagnostic horizon in a *Typic Aquisalids* profile in the coastal area

3.4 Diagnostic Soil Characteristics

Diagnostic soil characteristics are features of the soil that are used in various places in the keys or in the definitions of diagnostic horizons. In the soils of Kuwait, following are the diagnostic soil characteristics.

3.4.1 Free Carbonates

The term "free carbonates" is used as a criterion for the isotic mineralogy (no free carbonates) class. It refers to soil carbonates that are uncoated or unbound and that effervesce visibly or audibly when treated with cold, dilute HCl. The term "free carbonates" is nearly synonymous with the term "calcareous." Soils that have free carbonates generally have calcium carbonate as a common mineral, although sodium and magnesium carbonates are also included in this concept. Soils or horizons with free carbonates may have inherited the carbonate compounds from parent materials without any translocation or transformation processes acting on them. There is no implication of pedogenesis in the concept of free carbonates, as there is in identifiable secondary carbonates (defined below), although most forms of secondary carbonates are freely effervescent.

3.4.2 Identifiable Secondary Carbonates

The term "identifiable secondary carbonates" is used in the definitions of a number of taxa. It refers to translocated authigenic calcium carbonate that has been precipitated in place from the soil solution rather than inherited from a soil parent material, such as calcareous loess or limestone residuum.

Identifiable secondary carbonates either may disrupt the soil structure or fabric, forming masses, nodules, concretions, or spheroidal aggregates (white eyes) that are soft and powdery when dry, or may be present as coatings in pores, on structural faces, or on the undersides of rock or pararock fragments. If present as coatings, the secondary carbonates cover a significant part of the surfaces. Commonly, they coat all of the surfaces to a thickness of 1 mm or more. If little calcium carbonate is present in the soil, however, the surfaces may be only partially coated. The coatings must be thick enough to be visible when moist. Some horizons are entirely engulfed by carbonates. The color of these horizons is largely determined by the carbonates. The carbonates in these horizons are within the concept of identifiable secondary carbonates.

The filaments commonly seen in a dry calcareous horizon are within the meaning of identifiable secondary carbonates if the filaments are thick enough to be visible when the soil is moist. Filaments commonly branch on structural faces.

3.4.3 Aquic Conditions

Soils with aquic (L. *aqua*, water) conditions are those that currently undergo continuous or periodic saturation and reduction. The presence of these conditions is indicated by redoximorphic features, except in artificially drained soils. Artificial drainage is defined here as the removal of free water from soils having aquic conditions by surface mounding, ditches, or subsurface tiles or the prevention of surface or ground water from reaching the soils by dams, levees, surface pumps, or other means. In these soils water table levels and/or their duration are changed significantly in connection with specific types of land use. Upon removal of the drainage practices, aquic conditions would return. In the keys, artificially drained soils are included with soils that have aquic conditions.

3.4.4 Lithic Contact

A lithic contact is the boundary between soil and a coherent underlying material. The underlying material must be virtually continuous within the limits of a pedon. Cracks that can be penetrated by roots are few, and their horizontal spacing is 10 cm or more. The underlying material must be sufficiently coherent when moist to make hand-digging with a spade impractical, although the material may be chipped or scraped with a spade. The material below a lithic contact must be in a strongly cemented or more cemented rupture-resistance class. Commonly, the material is indurated. The underlying material considered here does not include diagnostic soil horizons, such as a duripan or a petrocalcic horizon.

A lithic contact is diagnostic at the subgroup level if it is within 50 cm of the mineral soil surface. In the soils of Kuwait lithic contact is used as soil series differentiae, where it was identified between 50–100 cm depth (Salim series- Typic torripsamments, mixed, hyperthermic; 100–200 cm in other two soil series (Farridah-Typic torripsamments, mixed, hyperthermic, and Awjah soil series-Typic Haplocalcids, sandy, mixed, hyperthermic).

3.4.5 Soil Moisture Regimes

The term "soil moisture regime" refers to the presence or absence either of ground water or of water held at a tension of less than 1500 kPa in the soil or in specific horizons during periods of the year. Water held at a tension of 1500 kPa or more is not available to keep most mesophytic plants alive. The availability of water is also affected by dissolved salts. If a soil is saturated with water that is too salty to be available to most plants, it is considered salty rather than dry. Consequently, a horizon is considered dry when the moisture tension is 1500 kPa or more and is considered moist if water is held at a tension of less than 1500 kPa but more than zero. A soil may be continuously moist in some or all horizons either throughout the year or for some part of the year. It may be either moist in winter and dry in summer or the reverse. In the Northern Hemisphere, summer refers to June, July, and August and winter refers to December, January, and February.

Normal Years

A normal year is defined as a year that has:

1. Annual precipitation that is plus or minus one standard deviation of the long-term (30 years or more) mean annual precipitation; *and*
2. Mean monthly precipitation that is plus or minus one standard deviation of the long-term monthly precipitation for 8 of the 12 months.

For the most part, normal years can be calculated from the mean annual precipitation; however, when catastrophic events occur during a year, the standard deviations of the monthly means should also be calculated. When precipitation data are evaluated to determine if the criterion for the presence of aquic conditions, or number of days that the moisture control section is moist, or number of days that some part of the soil is saturated has been met, it is permissible to include data from periods with below normal rainfall. Similarly, when precipitation data are evaluated to determine if the criterion for the number of days that the moisture control section is dry has been met, it is permissible to include data from periods with above normal rainfall. It is assumed that if the criteria are met during these periods, they will also be met during normal years.

3.4.5.1 Soil Moisture Control Section

The intent in defining the soil moisture control section is to facilitate estimation of soil moisture regimes from climatic data. The upper boundary of this control section is the depth to which a dry (tension of more than 1500 kPa, but not air-dry) soil will be moistened by 2.5 cm of water within 24 h. The lower boundary is the depth to which a dry soil will be moistened by 7.5 cm of water within 48 h. These depths do not include the depth of moistening along any cracks or animal burrows that are open to the surface.

If 7.5 cm of water moistens the soil to a densic, lithic, paralithic, or petroferric contact or to a petrocalcic or petrogypsic horizon or a duripan, the contact or the upper boundary of the cemented horizon constitutes the lower boundary of the soil moisture control section. If a soil is moistened to one of these contacts or horizons by 2.5 cm of water, the soil moisture control section is the boundary of the contact itself. The control section of such a soil is considered moist if the contact or upper boundary of the cemented horizon has a thin film of water. If that upper boundary is dry, the control section is considered dry.

The moisture control section of a soil extends approximately.

(1) from 10 to 30 cm below the soil surface if the particle-size class of the soil is fine-loamy, coarse-silty, fine-silty, or clayey;
(2) from 20 to 60 cm if the particle-size class is coarse-loamy; and
(3) from 30 to 90 cm if the particle-size class is sandy.

If the soil contains rock and pararock fragments that do not absorb and release water, the limits of the moisture control section are deeper. The limits of the soil moisture control section are affected not only by the particle-size class but also by differences in soil structure or pore-size distribution or by other factors that influence the movement and retention of water in the soil.

3.4.5.2 Classes Soil Moisture Regimes

The soil moisture regimes are defined in terms of the level of ground water and in terms of the seasonal presence or absence of water held at a tension of less than 1500 kPa in the moisture control section. It is assumed in the definitions that the soil supports whatever vegetation it is capable of supporting, i.e., crops, grass, or native vegetation, and that the amount of stored moisture is not being increased by irrigation or fallowing. These cultural practices affect the soil moisture conditions as long as they are continued.

Only two soil moisture regimes are recognized in Kuwait, Aquic and Aridic (or torric).

Aquic soil moisture regime.—The aquic (L. *aqua*, water) soil moisture regime is a reducing regime in a soil that is virtually free of dissolved oxygen because it is saturated by water. Some soils are saturated with water at times while dissolved oxygen is present, either because the water is moving or because the environment is unfavorable for micro-organisms (e.g., if the temperature is less than 1 °C); such a regime is not considered aquic.

It is not known how long a soil must be saturated before it is said to have an aquic soil moisture regime, but the duration must be at least a few days, because it is implicit in the concept that dissolved oxygen is virtually absent. Because dissolved oxygen is removed from ground water by respiration of microorganisms, roots, and soil fauna, it is also implicit in the concept that the soil temperature is above biologic zero for some time while the soil is saturated. Biologic zero is defined as 5 °C in this taxonomy.

Aridic and torric (L. *aridus*, dry, and L. *torridus*, hot and dry) soil moisture regimes.—These terms are used for the same moisture regime but in different categories of the taxonomy. In the aridic (torric) soil moisture regime, the moisture control section is, in normal years:

1. Dry in all parts for more than half of the cumulative days per year when the soil temperature at a depth of 50 cm below the soil surface is above 5 °C; *and*
2. Moist in some or all parts for less than 90 consecutive days when the soil temperature at a depth of 50 cm below the soil surface is above 8 °C.

Soils that have an aridic (torric) soil moisture regime normally occur in areas of arid climates. A few are in areas of semiarid climates and either have physical properties that keep them dry, such as a crusty surface that virtually precludes the infiltration of water, or are on steep slopes where runoff is high. There is little or no leaching in this soil moisture regime, and soluble salts accumulate in the soils if there is a source.

3.4.6 Soil Temperature Regimes

Following is a description of the soil temperature regime identified in Kuwait, which is described as "hyperthermic" and used at the soil family level of Kuwait soil taxonomy hierarchy.

Hyperthermic.—The mean annual soil temperature is 22 °C or higher, and the difference between mean summer and mean winter soil temperatures is 6 °C or more either at a depth of 50 cm below the soil surface or at a densic, lithic contact, whichever is shallower.

References

Gile LH, Peterson FF, Grossman RB (1966) Morphological and genetic sequences of carbonate accumulation in desert soils. Soil Sci 101:347–360

KISR (1999a) Soil survey for the State of Kuwait: Reconnaissance survey, vol II & III. Kuwait Institute for Scientific Research, Kuwait

KISR (1999b) Soil survey for the State of Kuwait: semi-detailed survey, vol IV & V. Kuwait Institute for Scientific Research, Kuwait

Omar SAS, Shahid SA (2013) Reconnaissance Soil Survey for the State of Kuwait. In: Shahid SA, Taha FK, Abdelfattah MA (eds) Chapter 3: in developments in soil classification, land use planning and policy implications: innovative thinking of soil inventory for land use planning and management of land resources, pp 85–110

Shahid SA, Abdelfattah MA, Wilson M (2007) A unique anhydrite soil in Abu Dhabi. United Arab Emirates. Soil Survey Horizons 48(4):75–79

Shahid SA, Abdelfattah MA, Wilson MA, Kelley JA, Chiaretti JV (2014) United Arab Emirates Keys to Soil Taxonomy. Springer, 108p

Soil Science Division Staff (2017) Soil survey manual. United States Department of Agriculture Handbook No. 18

Soil Survey Staff (2014) Keys to soil taxonomy, 12th edn. US Department of Agriculture, Natural Resources Conservation Service, US Government Printing Office, Washington, DC

Wilson MA, Shahid SA, Abdelfattah MA, Kelley JA, Thomas JE (2013) Anhydrite formation on the coastal sabkha of Abu Dhabi, United Arab Emirates. In: Shahid SA, Taha FK, Abdelfattah MA (eds) Chapter 8: developments in soil classification, land use planning and policy implications - innovative thinking of soil inventory for land use planning and management of land resources. Springer, pp 175–202

Atriplex leucoclada is commonly found in calcareous sandy soil

Families and Series Differentiae

4

Abstract

This chapter describes families and series differentiae used to classify the soils of Kuwait. The families and series are established using various criteria for mineral soils. The particle-size classes, mineralogy classes, cation-exchange activity classes, soil temperature class, soil depth classes of the control section of soil taxa are used. Particle-size classes (coarse-gypseous, sandy-skeletal, sandy, loamy, coarse-loamy, fine-loamy), Mineralogy classes (hypergypsic, gypsic, carbonatic and mixed), Cation-exchange activity classes (active, semiactive, subactive), Soil temperature class (hyperthermic) and Soil depth class (shallow) and lithic contact is used. The series criteria were used from 1–2 m depths for the occurrence of sandy, gypsic horizon, sodic (SAR > 13 $(mmoles/L)^{0.5}$, calcic horizon (0–1 m), coarse-loamy or fine-loamy, gypsic horizon, coarse-loamy and gypsic horizon, coarse-loamy and calcic horizon, fine-loamy and gypsic horizon on sandstone, sandy, gypsic horizon or petrogypsic horizon, sandy and lithic contact, sandy or coarse-loamy, petrocalcic horizon, and lithic contact (50–100 cm).

Keywords

Control section · Soil family · Soil series · Particle-size classes · Soil depth

S. A. Shahid and S. A. S. Omar, *Kuwait Soil Taxonomy*,
https://doi.org/10.1007/978-3-030-95297-6_4

4.1 Introduction

The following differentiae are used to distinguish families/series of mineral soils during the soil survey of Kuwait (KISR 1999a, b) using the guidelines of USDA-NRCS (Soil Survey Staff 1999), which is upgraded to the latest edition of Keys to Soil Taxonomy (Soil Survey Staff 2014). The class names of these differentiae are used to form the family name. The class names are listed and defined in the same sequence in which they appear in the family names. The names of the soil family are separated by comma (Typic Haplocalcids, sandy, mixed, hyperthermic). Exception is in the soil family with Typic Torripsamments subgroup, where particle size class is redundant, because Typic Torripsamments (by definition) are sandy. If more than one phase is needed to be expressed, they are separated by hyphen (Typic torripsamments, mixed, hyperthermic, shallow-lithic).

Particle-size classes and their substitutes
Mineralogy classes
Cation-exchange activity classes
Soil temperature classes
Soil depth classes
Phases of soil taxa

4.2 Definition of Particle-Size Classes

The term particle-size class is used to characterize the grain-size composition of the whole soil, including both the fine earth and the rock and pararock fragments up to the size of a pedon, but it excludes organic matter and salts more soluble than gypsum. The particle- size "texture" classes in pedologic classifications have been based on percentages, by weight, in the fraction less than 2.0 mm in diameter. Fine earth refers to particles smaller than 2 mm in diameter.

4.2.1 Control Section for Particle-Size Classes

The particle-size and substitute class names listed below are applied to certain horizons, or to the soil materials within specific depth limits, that have been designated as the control section for particle-size classes and their substitutes. The lower boundary of the control section may be at a specified depth (in centimeters) below the mineral soil surface or it may coincide with the upper boundary of a root-limiting layer (defined below).

4.2.1.1 Root-Limiting Layers

The concept of root-limiting layers as used in the Kuwait Soil Taxonomy defines the base of the soil horizons (a petrocalcic horizon, a petrogypsic horizon, a lithic contact) considered for most differentiae at the family level.

4.2.1.2 Key to the Control Section for Particle-Size Classes and Their Substitutes

The following list of particle-size control sections for the soils of Kuwait is arranged as a key to allow the user of the Kuwait Soil Taxonomy to make the correct classification by going through the key systematically, starting at the beginning and eliminating one by one all classes that include criteria that do not fit the soil in question. The soil belongs to the first class for which it meets all of the criteria listed.

A. For mineral soils that have a root-limiting layer (listed above) within 36 cm of the mineral soil surface, from the mineral soil surface to the root-limiting layer; *or*

B. Between the lower boundary of an Ap horizon or a depth of 25 cm below the mineral soil surface, whichever is deeper, and the shallower of the following: (a) a depth of 100 cm below the mineral soil surface or (b) a root-limiting layer.

4.2.1.3 Key to the Particle-Size and Substitute Classes

The key to the soils of Kuwait is arranged to allow the user of Kuwait soil taxonomy to make the correct particle-size distribution classification by going through the key systematically, starting at the beginning and eliminating one by one all classes that include criteria that do not fit the soil or layer in question. The class or substitute name of each layer within the control section must be determined from the key. If any two layers meet the criteria for strongly contrasting particle-size classes, the soil is named for that strongly contrasting class. If the soil has none of the strongly contrasting classes, the weighted average soil materials within the particle-size control section generally determine the class. Exceptions are soils that are not strongly contrasting and that have a substitute class name for one or more parts of the control section. In these soils the class or substitute name of the thickest (cumulative) part within the control section is used to determine the family name.

The sandy-skeletal, coarse-gypseous, sandy, loamy, coarse-loamy, fine-loamy particle-size and substitute classes have been identified in the soils of Kuwait. The gypseous-skeletal and fine-gypseous substitute classes are not recorded in the soil survey for the state of Kuwait, but are likely to be recorded in future soil survey work in Kuwait, therefore, these are also added in the following section.

Throughout the control section, meet one of the following particle-size class criteria.

1. The mineral; soils have, in the fraction less than 20 mm in diameter, 40 percent or more (by weight) gypsum *and* one of the following:

    a. A total of 35 percent or more (by volume) rock fragments.

    Gypseous-skeletal

    or

    b. Less than 35 percent (by volume) rock fragments and 50 percent or more (by weight) particles with diameter of 0.1 to 2.0 mm.

    Coarse-gypseous

    or

    c. Less than 35 percent (by volume) rock fragments.

    Fine-gypseous

2. The mineral soils that have a total content of rock fragments, plus any artifacts 2 mm or larger in diameter which are both cohesive and persistent, of 35 percent or more (by volume) *and* a texture class of coarse sand, sand, fine sand, loamy coarse sand, loamy sand, or loamy fine sand in the fine-earth fraction (Fig. 4.1).

    Sandy-skeletal

3. Have a texture class of coarse sand, sand, fine sand, loamy coarse sand, loamy sand, or loamy fine sand in the fine-earth fraction (Fig. 4.2).

    Sandy

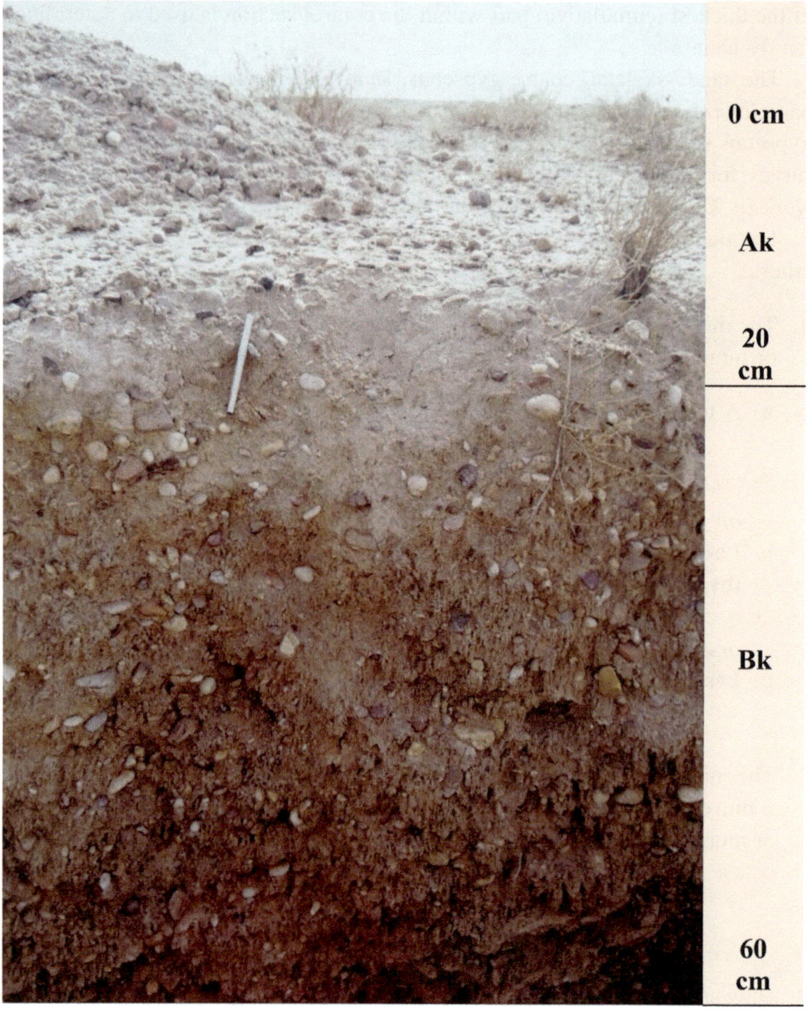

Fig. 4.1 Sandy-skeletal soil particle-size class

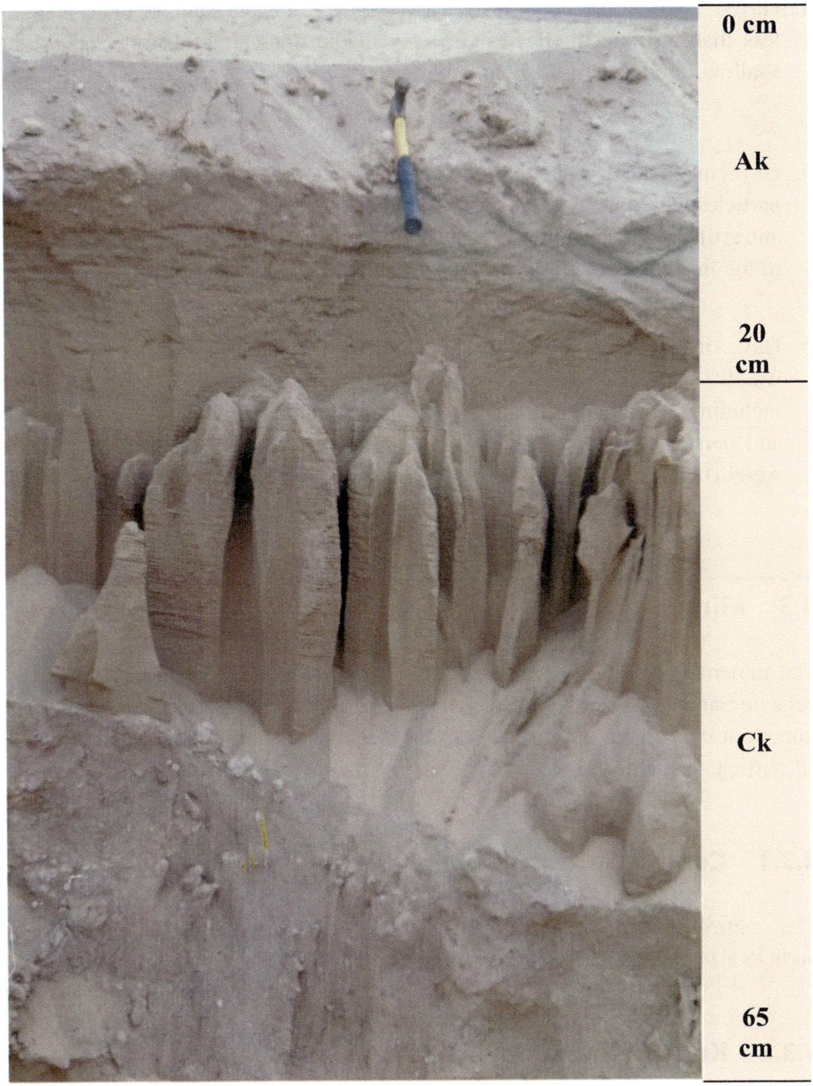

Fig. 4.2 Sandy soil texture in a *Typic torripsamments*

4. Have a texture class of loamy very fine sand, very fine sand, or finer, including less than 35 percent (by weight) clay in the fine-earth fraction, and are in a shallow family or in a Lithic subgroup.

Loamy

5. Have, in the fraction less than 75 mm in diameter, 15 percent or more (by weight) particles with diameters of 0.1 to 75 mm (fine sand or coarser, including gravel and artifacts 2 to 75 mm in diameter which are both cohesive and persistent) *and*, in the fine-earth fraction, less than 18 percent (by weight) clay.

Coarse-loamy

6. Have, in the fraction less than 75 mm in diameter, 15 percent or more (by weight) particles with diameters of 0.1 to 75 mm (fine sand or coarser, including gravel and artifacts 2 to 75 mm in diameter which are both cohesive and persistent) *and*, in the fine-earth fraction, 18 to less than 35 percent (by weight) clay.

Fine-loamy

4.3 Mineralogy Classes

The mineralogy of soils is known to be useful in making predictions about soil behavior and responses to management. Some mineralogy classes occur or are important only in certain taxa or particle-size classes, and others are important in all particle-size classes.

4.3.1 Control Section for Mineralogy Classes

The control section for mineralogy classes is the same as that defined for the particle-size classes and their substitutes.

4.3.2 Key to Mineralogy Classes

This key, like other keys in this Kuwait Soil Taxonomy, is designed in such a way that the reader makes the correct classification by going through the key systematically, starting at the beginning and eliminating one by one any classes that include criteria that do not fit the soil in question. The soil belongs to the first class for which it meets all of the required criteria. The user should first check the criteria in the first class and, if the soil in question does not meet the criteria listed there, proceed on to the next class, until the soil meets the criteria listed. All criteria are based on a weighted average.

The soils with horizons in the mineralogy control section that have a substitute class that replaces the particle-size class, other than fragmental, and that have:

1. 40 percent or more (by weight) gypsum either in the fine-earth fraction or in the fraction less than 20 mm in diameter, whichever has a higher percentage of gypsum.

Hypergypsic

or

2. Soil layers or horizons that have any particle-size class and 15 percent or more (by weight) gypsum, either in the fine-earth fraction (<2 mm) or in the fraction less than 20 mm in diameter, whichever has a higher percentage of gypsum (Fig. 4.3).

Gypsic

or

3. Soil layers or horizons that have any particle-size class and more than 40 percent (by weight) carbonates (expressed as $CaCO_3$) plus gypsum, either in the fine-earth fraction (<2mm) or in the fraction less than 20 mm in diameter, whichever has a higher percentage of carbonates plus gypsum.

Carbonatic

or

4. Soils with mineralogy classes other than hypergypsic, gypsic and carbonatic.

Mixed

4.4 Cation-Exchange Activity Classes

The cation-exchange activity classes help in making interpretations about the nutrient-holding capacity of soils and their suites of colloids. The cation-exchange capacity is determined by NH_4OAc at pH 7 on the fine-earth fraction (<2 mm). The Cation Exchange Capacity (CEC) of the organic matter, sand, silt, and clay is included in the determination.

Fig. 4.3 A profile of *Leptic Haplogypsids* showing Gypsic soil mineralogy class

The criteria for the classes use ratios of CEC to the percent, by weight, of silicate clay, calculated by weighted average in the control section. In the following classes "clay" excludes clay-size carbonates. Percent carbonate clay must be subtracted from percent total clay before calculating the CEC to clay ratio. If the ratio of percent water retained at 1500 kPa tension to the percentage of measured clay is 0.25 or less or 0.6 or more in half or more of the particle-size control section (or in a part of contrasting families), then the percentage of clay is estimated by the following formula: Clay % = 2.5 (% water retained at 1500 kPa tension - % organic carbon).

4.4.1 Use of the Cation-Exchange Activity Classes

The cation-exchange activity classes are used for soils classified in the mixed mineralogy class of coarse-loamy, loamy, particle-size classes. Cation-exchange activity classes are not assigned to Psamments, "psamm" great groups of Entisols, or other soils with sandy or sandy-skeletal particle size classes because the low clay content causes cation-exchange activity classes to be less useful and less reliable.

4.4.2 Control Section for Cation-Exchange Activity Classes

The control section for cation-exchange activity classes is the same as that used to determine the particle-size and mineralogy classes.

4.4.3 Key to Cation-Exchange Activity Classes

A. Soils that are not Psamments of Entisols, and that are not in a sandy or sandy-skeletal particle-size class or any substitute for a particle-size class throughout the entire control section, and that have:

1. A mixed mineralogy class; *and*
2. A ratio of cation-exchange capacity (by 1N NH_4OAc pH 7) to percent clay (by weight) of:

a. 0.60 or more

Superactive

or

b. 0.40 to 0.60

Active

or

c. 0.24 to 0.40

Semiactive

or

d. Less than 0.24

Subactive

4.5 Soil Temperature Class

Soil temperature class, as named and defined here, is used as part of the family name in soil survey of Kuwait. The Celsius (centigrade) scale is the standard. It is assumed that the temperature is that of a soil that is not being irrigated.

4.5.1 Control Section for Soil Temperature

The control section for soil temperature is either at a depth of 50 cm below the soil surface or at the upper boundary of a root-limiting layer, whichever is shallower. The soil temperature class is defined in terms of the mean annual soil temperature and the difference between mean summer and mean winter temperatures.

4.5.2 Key to Soil Temperature Class

Hyperthermic is the only soil temperature class identified in Kuwait. The soils that have a difference in soil temperature of 6 °C or more between mean summer (June, July, and August) and mean winter (December, January, and February) and a mean annual soil temperature of 22 °C (72 °F) or higher.

Hyperthermic

4.6 Soil Depth Classes

Soil depth class is used in families of Kuwait mineral soils that have a root-limiting layer at a specified depth from the mineral soil surface. The root-limiting layers included in soil depth classes are petrocalcic and petrogypsic horizons and lithic contact. One soil depth class name, "shallow," is used to characterize certain soil families that have one of the depths indicated in the following key.

4.6.1 Key to Soil Depth Classes

Mineral soils that are less than 50 cm deep (from the mineral soil surface) to a root-limiting layer (petrocalcic & petrogypsic horizons) and are not in a Lithic subgroup.

4.7 Series Differentiae Within a Family

The criteria most commonly used include presence of, depth to, thickness of, and expression of horizons and properties diagnostic for the higher categories and differences in texture, mineralogy, soil moisture, soil temperature, and amounts of organic matter. The limits of the properties used as differentiae must be more narrowly defined than the limits for the family. The properties used, however, must be reliably observable or be inferable from other soil properties or from the setting or vegetation. The differentiae used must be within the series control section. Differences in soil that are outside the series control section and that have not been recognized as series differentiae but are relevant to potential uses of certain soils are considered as a basis for ***phase*** distinctions.

4.7.1 Control Section for the Differentiation of Series

The control section for the soil series is similar to that for the family, but it differs in a few important respects. The particle-size and mineralogy control sections for families end at the upper boundary of certain diagnostic subsurface horizons, such as a petrocalcic horizon, petrogypsic horizon because these horizons have few roots. The thickness of such root-limiting horizons is taken into account in differentiating concepts of competing soil series, when they occur within the series control section. The series control section includes materials starting at the soil surface and extends into the first 25 cm of densic materials, a manufactured layer, or paralithic materials, if the densic, manufactured layer, or paralithic contacts, respectively, are less than 125 cm below the mineral soil surface. In contrast, the properties of materials below any densic, lithic, manufactured layer, paralithic, or petroferric contact are not used for classification in the categories above the series (i.e., order through family). The properties of horizons and layers below the particle-size control section, a depth between 100 and 150 cm (or to 200 cm if in a diagnostic horizon) from the mineral soil surface, also are considered in the series category of this taxonomy.

4.7.1.1 Key to the Control Section for the Differentiation of Series

The part of a soil to be considered in differentiating series within a family is as follows:
Mineral soils: From the soil surface to the shallowest of the following.

1. A Lithic contact; *or*
2. A depth of either 25 cm below a densic, manufactured layer, or paralithic contact or 150 cm below the soil surface, whichever is shallower, if there is a densic, manufactured layer, or paralithic contact within 150 cm; *or*

3. A depth of 150 cm if the base of the deepest diagnostic horizon is less than 150 cm from the soil surface; *or*
4. The lower boundary of the deepest diagnostic horizon or a depth of 200 cm, whichever is shallower, if the lower boundary of the deepest diagnostic horizon is 150 cm or more below the soil surface.

In the semi-detailed soil survey of Kuwait (KISR 1999b), series criteria were used from 1–2 m depths. In most the soil series identified in Kuwait, family criteria were sufficient to establish soil series. In others, following series criteria (1–2 m) were used, except where otherwise stated:

- Sandy, gypsic horizon
- Sodic [SAR > 13 $(\text{mmoles/l})^{0.5}$]
- Calcic horizon (0–1 m)
- Coarse-loamy or fine loamy, gypsic horizon
- Coarse-loamy, gypsic horizon
- Coarse-loamy, calcic horizon
- Fine-loamy, gypsic horizon
- On sandstone
- Sandy, gypsic horizon or petrogypsic horizon
- Sandy, lithic contact
- Sandy or coarse-loamy, petrocalcic horizon
- Sandy
- Lithic contact (50–100 cm)

More details about the soil series taxa and series differentia criteria can be found in Chap. 6.

References

KISR (1999a) Soil survey for the state of Kuwait: reconnaissance survey, vol II & III. Kuwait Institute for Scientific Research, Kuwait

KISR (1999b) Soil survey for the state of Kuwait: semi-detailed survey, vol IV & V. Kuwait Institute for Scientific Research, Kuwait

Soil Survey Staff (1999) Soil taxonomy: a basic system of soil classification for making and interpreting soil surveys, 2nd edn. US Department of Agriculture Handbook 436. Natural Resources Conservation Service

Soil Survey Staff (2014) Keys to soil taxonomy, 12th edn. US Department of Agriculture, Natural Resources Conservation Service, US Government Printing Office, Washington, DC

Citrullus colocynthis grows on sandy or silty ground, particularly in wadis and depressions

Identification of the Taxonomic Class of a Soil

5

Abstract

In this chapter, key to soil taxonomy hierarchy of Kuwait, from order-suborder-great group-subgroup are described. At the higher level of soil taxonomy, of 12 soil orders reported in US keys to soil taxonomy, only two soil orders (Aridisols and Entisols) have been identified in Kuwait. Entisols cover 30% and Aridisols cover 70% of the area surveyed. Within the soil order Aridisols, the soil taxa have been described as suborders, great groups and subgroups, such as, Suborders (Argids, Calcids, Cambids, Gypsids, Salids), Great Groups (Calciargids, Gypsiargids, Haplargids, Haplocalcids, Petrocalcids, Haplocambids, Argigypsids, Calcigypsids, Haplogypsids, Petrogypsids, Aquisalids), and Subgroups (Arenic Calciargids, Typic Calciargids, Typic Gypsiargids, Typic Haplargids, Typic Haplocalcids, Calcic Petrocalcids, Typic Petrocalcids, Typic Haplocambids, Calcic Salidic Argigypsids, Calcic Argigypsids, Lithic Calcigypsids, Typic Calcigypsids, Leptic Salidic Haplogypsids, Leptic Haplogypsids, Typic Haplogypsids, Petrocalcic Petrogypsids, Calcic Salidic Petrogypsids, Calcic Petrogypsids, Typic Petrogypsids, Gypsic Aquisalids, Typic Aquisalids). Haplosalids are not recorded in the soil survey for the state of Kuwait, but are likely to be recorded in future soil survey work in Kuwait, therefore, these are also added in this chapter. Within the order Entisols, the soil taxa have been described as Suborders (Orthents, Psamments), Great Groups (Torriorthents, Torripsamments), Subgroups (Salidic Torriorthents, Typic Torriorthents, Lithic Torripsamments, and

S. A. Shahid and S. A. S. Omar, *Kuwait Soil Taxonomy*,
https://doi.org/10.1007/978-3-030-95297-6_5

Typic Torripsamments). The key is only application to Aridisols and Entisols identified in Kuwait. Other countries in the GCC can develop their own national soil taxonomy books specific to soils identified in national surveys.

Keywords
Aridisols · Entisols · Argids · Calcids · Cambids · Gypsids · Salids · Taxonomic

5.1 Introduction

The soil survey of Kuwait (KISR 1999a, b) was completed at two levels, (i) Reconnaissance survey at family level of US Keys to Soil Taxonomy hierarchy and; (ii) Semi-detailed survey at soil series level (Soil Survey Staff 1999). During these two surveys, of twelve soil orders identified globally and described in the US Keys to Soil Taxonomy (Soil Survey Staff 2014) only two soil orders (Aridisols and Entisols) have been identified in the State of Kuwait. Therefore, Kuwait Soil Taxonomy is limited to soil taxa identified in Kuwait from Order-suborder-great group-subgroup-family level (KISR 1999a) and to Order-suborder-great group-subgroup-family-series level (KISR 1999b). The user of Kuwait Soil Taxonomy can assign a soil taxon to an in-depth described location by going through the key systematically, starting at the beginning and eliminating one by one all classes that include criteria that do not fit the soil in question until find the matched one. The key is only application to Aridisols and Entisols hierarchy identified in Kuwait. Other countries in the GCC can develop their own national soil taxonomy books based on the soils mapped within their national boundaries. Lately UAE has published United Arab Emirates Keys to Soil Taxonomy (Shahid et al. 2014) based on the results from three soil surveys (i) Dubai emirate (DM 2005), (ii) Soil Survey of Abu Dhabi Emirate (EAD 2009a, b) and Soil Survey of the Northern Emirates of UAE (EAD-MOEW 2012).

5.2 Soil Orders Identified in Kuwait

Two soil orders "*Aridisols & Entisols*" have been identified in Kuwait. According to the Soil Survey of Kuwait (KISR 1999a, b) Aridisols are dominant (70%) soils in Kuwait compared to Entisols (30%) (Fig. 5.1).

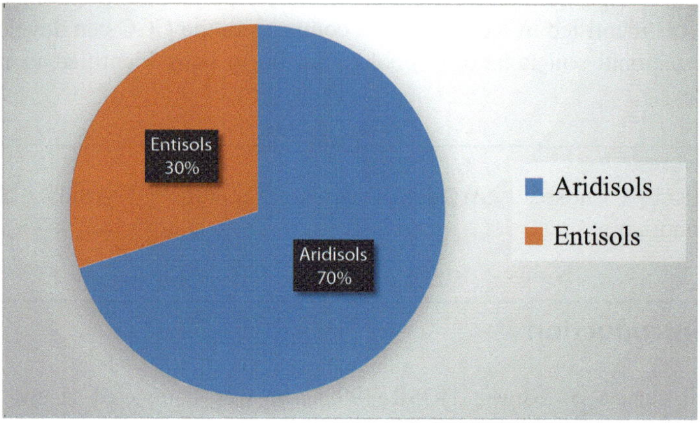

Fig. 5.1 Relative distribution of soil map units in Aridisols and Entisols

5.2.1 Aridisols

Aridisols, are soils that do not have water available to mesophytic plants for long periods. In hot climate soil moisture is held at potentials less than permanent wilting point or if it is salty, or both. There is no period of 90 consecutive days when moisture is continuously available for plant growth (Eswaran and Reich 2005).

Aridisols is one of the twelve soil orders in US Soil Taxonomy (Soil Survey Staff 2014). Aridisols (from the Latin *aridus*, for "dry", and *solum*) form in an arid or semi-arid climate. Aridisols dominate the deserts. The deserts occupy about one third of the Earth's land surface. Aridisols occupy approximately 12 percent of the Earth's ice-free land. Aridisols are low in organic matter, poor vegetation covers and are water deficient. Soil formation and development of diagnostic horizons (calcic, gypsic, argillic etc.) requires many years to develop. Accumulation of salts at surface is a common feature due to high evaporation and low leaching.

Fig. 5.2 A view of sandy desert landscape presenting Entisols soil order covering 30% of surveyed area in Kuwait

5.2.2 Entisols

In US Soil Taxonomy (Soil Survey Staff 2014), Entisols is one of the twelve soil orders, defined as soils that do not show any profile development other than an A horizon. All deep sandy soils (sand sheets, sand dunes) without any development are classified under Entisols. The Entisols has no diagnostic horizons (calcic, gypsic, salic, petrocalcic, petrogypsic etc.), and the parent material is unaltered. The Entisols are the most abundant soil order, occupying about 16% of the global ice-free land area (Fig. 5.2).

5.3 Understanding Soil Taxonomic Classes

It is mandatory that the user of this chapter for the identification of soil taxa to be well familiar with the definitions of master horizons, diagnostic horizons, and the terms defined in the Soil Survey Manual (Soil Science Division Staff 2017) and the Field Book for Describing and Sampling Soils (Schoeneberger et al. 2012).

To describe the key to soil orders, suborders, great groups and subgroups for the soils of Kuwait, the excerpt from US Keys to Soil Taxonomy (Soil Survey Staff 2014) relevant to the soils of Kuwait is extracted and described below.

5.4 Key to Soil Orders

The soils that:

1. Have:

   a. An aridic soil moisture regime; *and*
   b. An ochric epipedon; *and*
   c. *One or more* of the following within 100 cm of the soil surface: a cambic horizon with a lower depth of 25 cm or more; a calcic, gypsic, petrocalcic, petrogypsic, or salic horizon; *or*
   d. An argillic horizon; *or*

2. Have a salic horizon; *and*

   a. Saturation with water in one or more layers within 100 cm of the soil surface for 1 month or more during a normal year; *and*
   b. A moisture control section that is dry in some or all parts at some time during normal years.

Aridisols, p. 63

or

Other soils that do not classify as Aridisols.

Entisols, p. 75

5.5 Key to Suborders of Aridisols

It is to be noted that very few auger holes/profiles were described in the argids and cambids suborders during the soil survey of Kuwait. Therefore, due to their minor extent, they were not identified as separate soil families or series, and thus not included in the map units, however, they are included in Kuwait Soil Taxonomy.

Aridisols that have a salic horizon within 100 cm of the soil surface.

Salids, p. 72

Other Aridisols that have a gypsic or petrogypsic horizon within 100 cm of the soil surface and do not have a petrocalcic horizon overlying these horizons.

Gypsids, p. 67

Other Aridisols that have an argillic horizon within 100 cm of the soil surface.

Argids, p. 64

Other Aridisols that have a calcic or petrocalcic horizon within 100 cm of the soil surface.

Calcids, p. 65

Other Aridisols that have a cambic horizon within 100 cm of the soil surface.

Cambids, p. 67

5.5.1 Argids

Key to Great Groups

Argids that have a gypsic horizon within 150 cm of the soil surface.

Gypsiargids, p. 65

Other Argids that have a calcic horizon within 150 cm of the soil surface.

Calciargids, p. 64

Other Argids that have only argillic horizon within 150 cm of the soil surface and lack properties of other horizons.

Haplargids, p. 65

Calciargids

Key to Subgroups

Calciargids that have a texture class (fine-earth fraction) of coarse sand, sand, fine sand, loamy coarse sand, loamy sand, or loamy fine sand throughout a layer extending from the soil surface to the top of an argillic horizon at a depth of 50 cm or more.

Arenic Calciargids

Other Calciargids that represent the central concept of the great group.

Typic Calciargids

Gypsiargids

Key to Subgroups

The argids that have only a gypsic horizon within 150 cm of the soil surface. These soils represent the central concept of the great group.

Typic Gypsiargids

Haplargids

Key to Subgroups

Argids that have only argillic horizon within 150 cm of the soil surface and lack properties of other horizons. These soils represent central concept of the great group.

Typic Haplargids

5.5.2 Calcids

Key to Great Groups

Calcids that have a petrocalcic horizon within 100 cm of the soil surface.

Petrocalcids, p. 65

Other soils which are most extensive in the Calcids and do not present petrocalcic horizon within the upper 100 cm of the soil surface.

Haplocalcids, p. 65

Haplocalcids

Key to Subgroups

Haplocalcids that have a calcic horizon within 100 cm of the soil surface and lack all other characteristics of any other subsurface diagnostic horizons. These soils represent the central concept of the great group. (Fig. 5.3)

Typic Haplocalcids

Petrocalcids

Key to Subgroups

Petrocalcids that have a calcic horizon overlying the petrocalcic horizon.

Calcic Petrocalcids

Fig. 5.3 A soil profile of *Typic Haplocalcids*

Other Petrocalcids that have a petrocalcic horizon within 100 cm of the soil surface and lack all other characteristics of any other subsurface diagnostic horizon. These soils represent the central concept of the great group.

Typic Petrocalcids

5.5.3 Cambids

Key to Great Groups

Cambids that have only cambic horizon within 100 cm of the soil surface and lack all other characteristics of any other subsurface diagnostic horizon.

Haplocambids, p. 67

Haplocambids

Key to Subgroups

Haplocambids that have only cambic horizon within 100 cm of the soil surface and lack all other characteristics of any other subsurface diagnostic horizon. These soils represent the central concept of the great group.

Typic Haplocambids

5.5.4 Gypsids

Key to Great Groups

Gypsids that have a petrogypsic or petrocalcic horizon within 100 cm of the soil surface.

Petrogypsids, p. 70

Other Gypsids that have an argillic horizon within 100 cm of the soil surface.

Argigypsids, p. 68

Other Gypsids that have a calcic horizon within 100 cm of the soil surface.

Calcigypsids, p. 68

Other Gypsids that have only a gypsic horizon within 100 cm of the soil surface.

Haplogypsids, p. 68

Argigypsids

Key to Subgroups

Argigypsids that have both:

1. An ECe of more than 8 to less than 30 dS m^{-1} in a layer 10 cm or more thick, within 100 cm of the soil surface: *and.*
2. A calcic horizon overlying the gypsic horizon

Calcic Salidic Argigypsids

Other Argigypsids that have:
A calcic horizon overlying the gypsic horizon.

Calcic Argigypsids

Calcigypsids

Key to Subgroups

Calcigypsids that have a lithic contact within 50 cm of the soil surface.

Lithic Calcigypsids

Calcigypsids that lack all other diagnostic characteristics. These soils represent the central concept of the great group (Fig. 5.4).

Typic Calcigypsids

Haplogypsids

Key to Subgroups

Haplogypsids that have both:

1. An ECe of more than 8 to less than 30 dS m^{-1} in a layer 10 cm or more thick, within 100 cm of the soil surface: *and.*
2. A gypsic horizon within 18 cm of the soil surface.

Leptic Salidic Haplogypsids

Other Haplogypsids that have a gypsic horizon within 18 cm of the soil surface (Fig. 5.5).

Leptic Haplogypsids

Fig. 5.4 A profile of *Typic Calcigypsids*. Gypsum accumulation begins at 90 cm depth. A calcic horizon occurs at 30–90 cm depth. A decalcified layer at 0–30 cm

Other Haplogypsids that have only a gypsic horizon and lack all other characteristics. These soils represent the central concept of the great group.

Typic Haplogypsids

Fig. 5.5 A profile of *Leptic Haplogypsids*. Gypsum accumulation begins at 12–15 cm from soil surface

Petrogypsids

Key to Subgroups

Petrogypsids that have a petrocalcic horizon within 100 cm of the soil surface (Fig. 5.6).

Petrocalcic Petrogypsids

Other Petrogypsids that have both:

Fig. 5.6 A profile of *Petrocalcic Petrogypsids*. The horizon below 25 cm meets the criteria for both a petrocalcic and a petrogypsic horizon

1. An ECe of more than 8 to less than 30 dS m^{-1} in a layer 10 cm or more thick, within 100 cm of the soil surface: *and*
2. A calcic horizon overlying the petrogypsic horizon.

Calcic Salidic Petrogypsids

Other Petrogypsids that have a calcic horizon overlying the petrogypsic horizon (Fig. 5.7).

Calcic Petrogypsids

Other Petrogypsids that have only a petrogypsic horizon and lack all other diagnostic subsurface horizons or characteristics. These soils represent the central concept of the great group (Fig. 5.8).

Typic Petrogypsids

Fig. 5.7 A profile of *Calcic Petrogypsids*. A calcic horizon occurs at 20 cm and petrogypsic at 40 cm depth

5.5.5 Salids

Key to Great Groups

Salids that are saturated with water in one or more layers within 100 cm of the mineral soil surface for 1 month or more in normal years.

Aquisalids, p. 72

Other Salids that are not saturated with water in one or more layers within 100 cm of the mineral soil surface for 1 month or more in normal years.

Haplosalids, p. 75

Aquisalids

Key to Subgroups

Aquisalids that have a gypsic horizon within 100 cm of the soil surface.

Gypsic Aquisalids

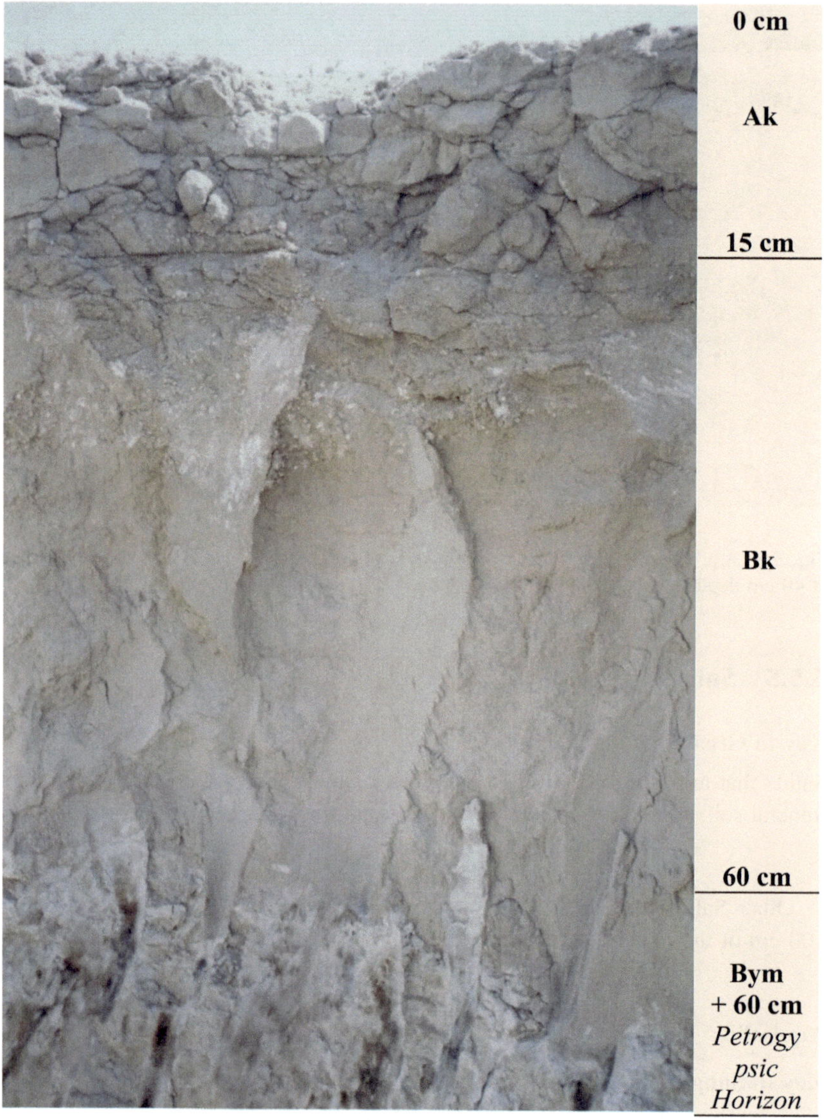

Fig. 5.8 A soil profile showing *Typic Petrogypsids* (Total depth 75 cm). Petrogypsic horizon started at 60 cm depth

Other Aquisalids that have only a salic horizon and do not have other subsurface diagnostic horizons within the 100 cm of the soil surface. These soils represent the central concept of the great group (Fig. 5.9).

Typic Aquisalids

Fig. 5.9 A soilscape showing *Typic Aquisalids* in the coastal area

Haplosalids

Key to Subgroups

Haplosalids that have only a salic horizon and do not have other subsurface diagnostic horizons within the 100 cm of the soil surface. These soils represent the central concept of the great group.

Typic Haplosalids

5.6 Key to Suborders of Entisols

Entisols that have less than 35 percent (by volume) rock fragments and a texture class of loamy fine sand or coarser in all layers (sandy loam lamellae are permitted) within the particle-size control section.

Psamments, p. 76

Other Entisols that occur on recent erosional surfaces. As such they typically represent soils that display a high content of rock fragments or shell fragments (exceeding 35% by volume) or a texture of loamy fine sand or finer in any horizon within the 100 cm of the soil surface. The soils are sandy or loamy in texture, are well drained.

Orthents, p. 75

5.6.1 Orthents

Key to Great Groups

Orthents that have an aridic (or torric) soil moisture regime.

Torriorthents, p. 76

Torriorthents

Key to Subgroups

Torriorthents that have:

1. An ECe of more than 8 to less than 30 dS m^{-1} in a layer 10 cm or more thick, within 100 cm of the soil surface: *and*

Salidic Torriorthents

Torriorthents that lack all other diagnostic characteristics. These are the soils that represent the central concept of the great group. They are typically moderately deep or deep, well drained, gravelly soils.

Typic Torriorthents

5.6.2 Psamments

Key to Great Groups

Psamments that have an aridic (or torric) soil moisture regime and do not have moisture available for plants for long periods.

Torripsamments, p. 76

Torripsamments

Key to Subgroups

Torripsamments that have a lithic contact within 50 cm of the soil surface.

Lithic Torripsamments

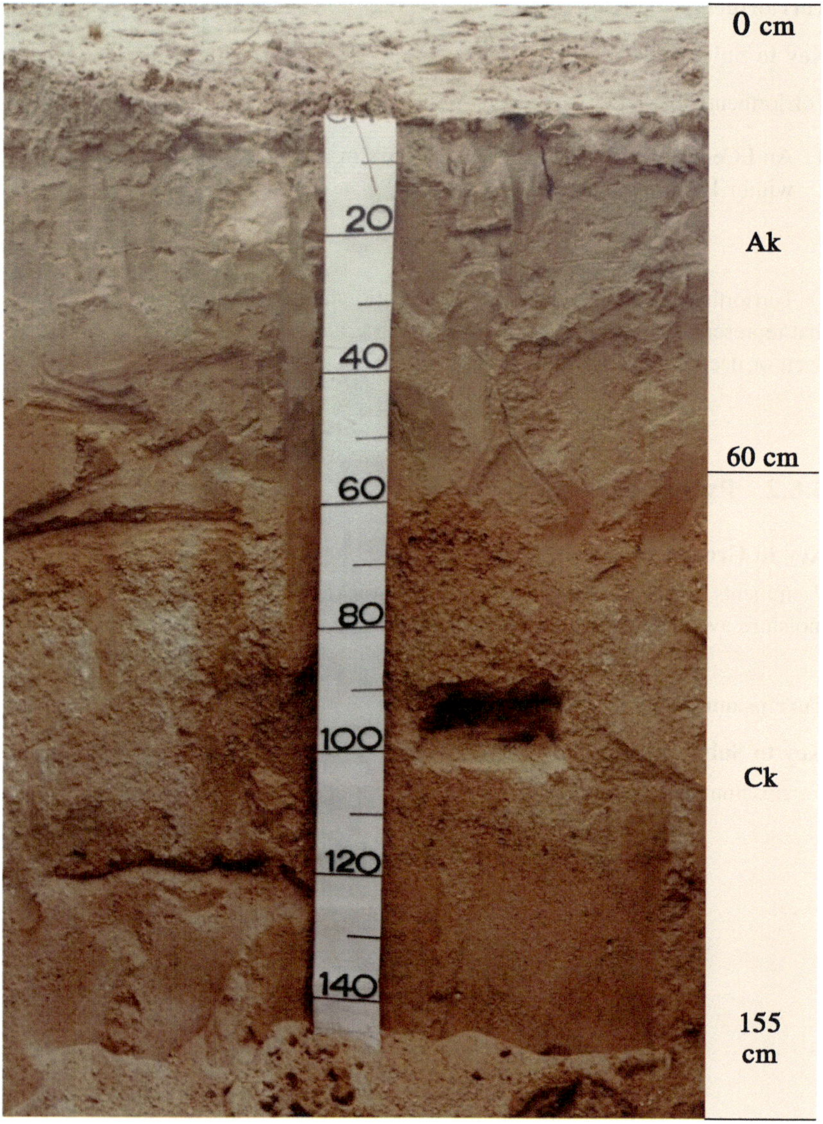

Fig. 5.10 A soil profile showing *Typic Torripsamments*

Torripsamments that lack all other diagnostic characteristics. These are the soils that represent the central concept of the great group. They are deep and have a texture of sand or loamy sand throughout the top 100 cm of the profile. They are nonsaline and are always calcareous to various degrees but do not have enough pedogenic carbonate accumulation to form a calcic horizon. Torripsamments (27.3%) are the second most extensive soils in Kuwait (Fig. 5.10) after petrogypsids (33.4%).

Typic Torripsamments

References

Dubai Municipality (DM) (2005) Satellite imagery and thematic mapping project, 221p. Executed jointly by Global Scan Technologies (Dubai) and National Remote Sensing Agency, India

EAD (2009a) Soil survey of Abu Dhabi Emirate-extensive survey, vol 1, pp xx+506. Environment Agency–Abu Dhabi, UAE

EAD (2009b) Soil survey of Abu Dhabi emirate, Intensive survey, vol III, pp xviii+435. Environment Agency, Abu Dhabi

EAD-MOEW (2012) Soil survey of the Northern Emirates. Soil Report, vol 1, pp xxvi+410. Environment Agency–Abu Dhabi, UAE

Eswaran H, Reich PF (2005) Aridisols–an overview. In: Encyclopedia of Soils in the Environment

KISR (1999a) Soil survey for the State of Kuwait. Reconnaissance survey, vol II. Kuwait Institute for Scientific Research, Kuwait

KISR (1999b) Soil survey for the State of Kuwait. Semi-detailed survey, vol IV. Kuwait Institute for Scientific Research, Kuwait

Schoeneberger PJ, Wysocki DA, Benham EC, Soil Survey Staff (2012) Field book for describing and sampling soils, version 3.0. Natural Resources Conservation Service, National Soil Survey Center

Shahid SA, Abdelfattah MA, Wilson MA, Kelley JA, Chiaretti JV (2014) United Arab emirates keys to soil taxonomy, 108p. Springer

Soil Science Division Staff (2017) Soil survey manual. U.S. Department of Agriculture Handbook 18. Soil Conservation Service

Soil Survey Staff (1999) Soil taxonomy: a basic system of soil classification for making and interpreting soil surveys, 2nd edn. US Department of Agriculture Handbook 436. Natural Resources Conservation Service

Soil Survey Staff (2014) Keys to soil taxonomy, 12th edn. US Department of Agriculture, Natural Resources Conservation Service, US Government Printing Office, Washington, DC

Silene villosa grows on sandy soils (Entisols)

Kuwait Soil Taxonomy Hierarchy Soil Families and Soil Series

6

Abstract

The soil taxonomy hierarchy of Kuwait is based on two soil surveys, (i) Fourth order Reconnaissance soil survey of Kuwait completed at 1:100,000 scale at the fifth hierarchy level (soil family) of US Soil Taxonomy; and (ii) Second order Semi-detained soil survey of 200,000 ha at scale 1:25,000 scale (with the potential for irrigated agriculture) completed at the sixth level (soil series) of US Soil taxonomy hierarchy. Based on the reconnaissance survey, 2 soil orders, 7 suborders, 8 great groups, 14 subgroups and 24 soil families were mapped. In the semi-detailed survey 39 soil series were mapped. Mixed, carbonatic, gypsic and hypergypsic mineralogy classes are identified. Sandy, sandy-skeletal, coarse-loamy, fine-loamy and coarse-gypseous family textural classes are identified. In this book the terminologies and the standards have been upgraded to the latest USDA protocols.

Keywords

Taxonomy · Hierarchy · Mineralogy class · Particle size class · Soil family · Soil series · Kuwait

6.1 Introduction

The US Soil Taxonomy was used to establish soil families (reconnaissance survey) and soil series (semi-detailed survey). The US Soil Taxonomy (Soil Survey Staff 2014) hierarchy constitutes six levels.

S. A. Shahid and S. A. S. Omar, *Kuwait Soil Taxonomy*,
https://doi.org/10.1007/978-3-030-95297-6_6

Soil Orders (2)
↓
Soil Suborders (7)
↓
Soil Great Groups (10)
↓
Soil Subgroups (18)
↓
Soil Families (24)
↓
Soil Series (39) Semi-detailed survey

Based on the reconnaissance and semi-detailed soil surveys (KISR 1999a, b), 2 soil orders, 7 suborders, 10 great groups, 18 subgroups, 24 soil families and 39 soil series were mapped. The name of soil series is given due to the location, where first time it is recognized. While developing soil maps in both surveys, soil families are the component soils of the map units in reconnaissance survey, and soil series as component soils of the map units in semi-detailed survey maps. Among the soil taxonomy hierarchy, soil series (6th level) are the most homogeneous classes and are the most common tool for communicating information about a particular soil type. As a general guide, a new soil series differs appreciably in either morphology or composition, or both, from already defined soil series. Differences in relevant characteristics must be larger than what may be normal errors of observation or estimates. Tables 6.1 and 6.2 provide details of the soil great groups and soil subgroups identified in reconnaissance and semi-detailed soil survey of Kuwait. The surveyors are recommended to key out soils in future surveys of Kuwait and to correlate the soil information with Tables 6.3 and 6.4, which are providing the taxonomic names of families and series. Further information about soil profiles (physical, chemical, mineralogical characteristics), soil map units and thematic maps can be obtained in soil survey reports (KISR 1999a, b).

To key out soil taxa at various levels. In this chapter only brief description is given with respect to soil taxonomy hierarchy of Kuwait.

6.2 Soil Orders

Among the 12 soil orders recognized globally, in Kuwait only two *(Aridisols and Entisols)* have been mapped.

Table 6.1 Great groups identified in reconnaissance (KISR 1999a) and semi-detailed survey (KISR 1999b) of Kuwait

Great groups Reconnaissance soil survey	Great groups Semi-detailed soil survey
Haplocalcids	Haplocalcids
Petrocalcids	Petrocalcids
Calcigypsids	Calcigypsids
Petrogypsids	Petrogypsids
Aquisalids	Aquisalids
Haplosalids	Haplosalids
Torripsamments	Torripsamments
Haplogypsids	Haplogypsids
Torriorthents	–
–	Argigypsids
–	Gypsiargids

Table 6.2 Soil subgroups identified in reconnaissance (KISR 1999a) and semi-detailed survey (KISR 1999b) of Kuwait

Soil subgroups—reconnaissance soil survey	Soil subgroups—semi-detailed soil survey
Typic Haplocacids	Typic Haplocalcids
Typic Petrocalcids	–
Calcic Petrocalcids	Calcic Petrocalcids
Typic Petrogypsids	Typic Petrogypsids
Calcic Petrogypsids	Calcic Petrogypsids
Petrocalic Petrogypsids	Petrocalic Petrogypsids
Petrocalcic Salidic Petrogypsids	–
Typic Calcigypsids	Typic Calcigypsids
–	Lithic Calcigypsids
Leptic Haplogypsids	Leptic Haplogypsids
Typic Haplogypsids	Typic Haplogypsids
–	Calcic Salidic Argigypsids
–	Typic Gypsiargids
Gypsic Aquisalids	Gypsic Aquisalids
Typic Aquisalids	–
Salidic Torriorthents	–
–	Lithic Torripsamments
Typic Torripsammnts	Typic Torripsammnts

Table 6.3 Soil families identified during the Reconnaissance Soil Survey for the State of Kuwait and correlated to Kuwait Soil Taxonomy (KISR 1999a)

Soil family no	Subgroup	Particle size class	Mineralogy class	Cation-exchange activity class	Temperature regime	Identifier
1	Calcic Petrocalcids	Sandy	Mixed		Hyperthermic	
2	Calcic Petrocalcids	Sandy	Mixed		Hyperthermic	Shallow
3	Calcic Petrocalcids	Coarse-loamy	Mixed	Subactive	Hyperthermic	
4	Calcic Salidic Petrogypsids	Sandy	Mixed		Hyperthermic	
5	Gypsic Aquisalids	Sandy	Mixed		Hyperthermic	
6	Leptic Haplogypsids	Coarse-gypseous	Hypergypsic		Hyperthermic	
7	Petrocalcic Petrogypsids	Sandy	Mixed		Hyperthermic	
8	Petrocalcic Salidic Petrogypsids	Sandy	Mixed		Hyperthermic	Shallow
9	Petrocalcic Salidic Petrogypsids	Coarse-loamy	Mixed	Active	Hyperthermic	
10	Typic Aquisalids	Sandy	Mixed		Hyperthermic	
11	Typic Calcigypsids	Sandy	Mixed		Hyperthermic	
12	Typic Calcigypsids	Coarse-gypseous	Hypergypsic		Hyperthermic	
13	Typic Calcigypsids	Coarse-loamy	Mixed	Semiactive	Hyperthermic	
14	Typic Haplocalcids	Sandy	Mixed		Hyperthermic	
15	Typic Haplocalcids	Coarse-loamy	Mixed	Semiactive	Hyperthermic	
16	Typic Haplogypsids	Sandy	Mixed		Hyperthermic	
17	Typic Haplogypsids	Coarse-gypseous	Hypergypsic		Hyperthermic	

(continued)

Table 6.3 (continued)

Soil family no	Subgroup	Particle size class	Mineralogy class	Cation-exchange activity class	Temperature regime	Identifier
18	Typic Haplogypsids	Coarse-loamy	Mixed	Semiactive	Hyperthermic	
19	Typic Petrocalcids	Sandy	Mixed		Hyperthermic	
20	Typic Petrocalcids	Sandy	Mixed		Hyperthermic	Shallow
21	Typic Petrogypsids	Coarse-gypseous	Hypergypsic		Hyperthermic	
22	Typic Petrogypsids	Coarse-gypseous	Hypergypsic		Hyperthermic	Shallow
23	Salidic Torriorthents	Sandy-skeletal	Carbonatic		Hyperthermic	
24	Typic Torripsamments		Mixed		Hyperthermic	

Table 6.4 Soil series identified during the Semi-Detailed Soil Survey for the State of Kuwait and correlated to Kuwait Soil Taxonomy (KISR 1999b)

Soil Series	Subgroup	Particle size class	Mineralogy class	Cation-exchange activity class	Temperature regime	Identifier	Series criteria (1–2 m)
Lulidah	Gypsic Aquisalids	Sandy	Mixed		Hyperthermic		Sandy, gypsic horizon, sodic SAR > 13, saline
Umm ar-Rimam	Petrocalcic Petrogypsids	Sandy	Mixed		Hyperthermic		Calcic horizon 0–1 m
Mityahah	Petrocalcic Salidic Petrogypsids	Sandy	Mixed		Hyperthermic	Shallow	Family criteria are sufficient
Naayim	Petrocalcic Salidic Petrogypsids	Coarse-loamy	Mixed	Active	Hyperthermic		Calcic horizon 0–1 m
Abatih	Petrocalcic Petrogypsids	Loamy	Mixed	Semiactive	Hyperthermic	Shallow	Family criteria are sufficient
Mujaylis	Calcic Petrogypsids	Sandy	Mixed		Hyperthermic		Family criteria are sufficient
Abu Rasayn	Typic Petrogypsids	Coarse-gypseous	Hypergypsic		Hyperthermic	Shallow	Family criteria are sufficient
Ruhayyah	Calcic Salidic Argigypsids	Fine-loamy	Mixed	Active	Hyperthermic		Coarse-loamy or fine-loamy
Abdali	Typic Gypsiargids	Fine-loamy	Mixed	Active	Hyperthermic		Fine loamy, gypsic horizon
Talhah	Lithic Calcigypsids	Sandy	Mixed		Hyperthermic		On sandstone. Lithic contact at 45–100 cm (R horizon)

(continued)

Table 6.4 (continued)

Soil Series	Subgroup	Particle size class	Mineralogy class	Cation-exchange activity class	Temperature regime	Identifier	Series criteria (1–2 m)
Maqwa	Typic Calcigypsids	Sandy	Mixed		Hyperthermic		Sandy, gypsic horizon
Rukham	Typic Calcigypsids	Sandy	Gypsic		Hyperthermic		Sandy, gypsic horizon
Sabriyah	Typic Calcigypsids	Coarse-loamy	Mixed	Semiactive	Hyperthermic		Sandy, gypsic horizon
Umm Qudayr	Typic Calcigypsids	Coarse-loamy	Mixed	Superactive	Hyperthermic		Coarse-loamy, gypsic horizon
Rawdatayn	Typic Calcigypsids	Fine-loamy	Mixed	Semiactive	Hyperthermic		Fine loamy or coarse-loamy, gypsic horizon
Bahrah	Leptic Haplogypsids	Sandy	Mixed		Hyperthermic		Sandy, gypsic horizon
Tarfawi	Leptic Salidic Haplogypsids	Coarse-gypseous	Hypergypsic		Hyperthermic		Sandy, gypsic horizon and/or petrogypsic horizon
Qashaniyah	Typic Haplogypsids	Sandy	Mixed		Hyperthermic		Sandy, gypsic horizon
Arfajiyah	Typic Haplogypsids	Coarse-gypseous	Hypergypsic		Hyperthermic		Sandy, gypsic horizon
Saddah	Calcic Petrocalcids	Sandy	Mixed		Hyperthermic		Family criteria are sufficient
Zaqlah	Calcic Petrocalcids	Sandy	Mixed		Hyperthermic	Shallow	Family criteria are sufficient

(continued)

Table 6.4 (continued)

Soil Series	Subgroup	Particle size class	Mineralogy class	Cation-exchange activity class	Temperature regime	Identifier	Series criteria (1–2 m)
Zahr al-Liyah	Calcic Petrocalcids	Coarse-loamy	Mixed	Subactive	Hyperthermic		Family criteria are sufficient
Radhah	Calcic Petrocalcids	Loamy	Mixed	Active	Hyperthermic	Shallow	Family criteria are sufficient
Awjah	Typic Haplocalcids	Sandy	Mixed		Hyperthermic		Sandy, lithic contact
Manaqish	Typic Haplocalcids	Sandy	Mixed		Hyperthermic		Sandy or coarse-loamy, petrocalcic horizon
Juwayhil	Typic Haplocalcids	Sandy	Mixed		Hyperthermic		Sandy, gypsic horizon
Sulaybiyah	Typic Haplocalcids	Sandy	Mixed		Hyperthermic		Sandy, calcic horizon
Bihayth	Typic Haplocalcids	Sandy	Mixed		Hyperthermic		Sandy
Umm al-Luhud	Typic Haplocalcids	Sandy	Mixed		Hyperthermic		Coarse-loamy or fine-loamy, gypsic horizon
Shiqq	Typic Haplocalcids	Coarse-loamy	Mixed	Superactive	Hyperthermic		Coarse-loamy, calcic horizon
Mirfi	Lithic Torripsamments		Mixed		Hyperthermic		On sandstone. Paralithic contact at 8–45 cm (Cr horizon); lithic contact at 45–75 cm (R horizon)

(continued)

Table 6.4 (continued)

Soil Series	Subgroup	Particle size class	Mineralogy class	Cation-exchange activity class	Temperature regime	Identifier	Series criteria (1–2 m)
Salim	Typic Torripsamments		Mixed		Hyperthermic		Lithic contact (50–100 cm)
Farridah	Typic Torripsamments		Mixed		Hyperthermic		Sandy, lithic contact (110–197 cm)
Sur	Typic Torripsamments		Mixed		Hyperthermic		Sandy or coarse-loamy, petrocalcic horizon (110–120 cm & 180–200 cm)
Umm Niqa	Typic Torripsamments		Mixed		Hyperthermic		Sandy, gypsic horizon
Fawaris	Typic Torripsamments		Mixed		Hyperthermic		Sandy, calcic horizon
Muslan	Typic Torripsamments		Mixed		Hyperthermic		Sandy
Kabd	Typic Torripsamments		Mixed		Hyperthermic		Coarse-loamy or less commonly fine-loamy or fine, gypsic horizon
Wafra	Typic Torripsamments		Mixed		Hyperthermic		Coarse-loamy, calcic horizon

6.2.1 Entisols

Sandy soils without any enrichment of $CaCO_3$ (secondary $CaCO_3$), gypsum, salts, gravels, hardpan (30% of survey area).

6.2.2 Aridisols

Soils with the accumulation of salts, $CaCO_3$, gypsum, clay, hardpan/gatch etc. (70% of survey area).

6.3 Soil Suborders

At the second level of soil taxonomy hierarchy seven soil suborders have been mapped.

Salids—the soils in the order *Aridisols* that have a salic horizon within 100 cm of the soil surface.
Gypsids—the soils in the order *Aridisols* that have a gypsic or petrogypsic horizon within 100 cm of the soil surface and do not have a petrocalcic horizon overlying these horizons.
Argids—the soils in the order *Aridisols* that have an argillic horizon and do not have a petrocalcic horizon within 100 cm of the soil surface.
Calcids—the soils in the order *Aridisols* that have a calcic or petrocalcic horizon within 100 cm of the soil surface.
Cambids—the soils in the order *Aridisols* that have a cambic horizon within 100 cm of the soil surface.
Psamments—the soils in the order *Entisols* that have less than 35 % (by volume) rock fragments and a texture class of loamy fine sand or coarser in all layers (sandy loam lamellae are permitted) within the particle-size control section.
Orthents—the soils in the order *Entisols* that occur on recent erosional surfaces. As such they typically represent soils that display a high content of rock fragment, sea shells etc. The soils are sandy or loamy in texture, are well drained.

6.4 Soil Great Groups

At the third level of soil taxonomy hierarchy ten soil great groups have been mapped in both soil surveys of Kuwait (KISR 1999a, b). Haplosalids are not recorded in the soil survey for the state of Kuwait, but are likely to be recorded in future soil survey work in Kuwait, therefore, these are also added in the following section and Table 6.1.

- **Haplocalcids**—the soils with calcic horizon in the upper 1 m.
- **Petrocalcids**—the soils with petrocalcic horizon in the upper 1 m.
- **Calcigypsids**—the gypsids (with gypsic horizon) soil that also have a calcic horizon within 100 cm of the soil surface.
- **Petrogypsids**—the soils that have gypsic hardpan (petrogypsic horizon) within upper 1 m.
- **Aquisalids**—the hypersaline (salic horizon) soils (coastal & inland sabkha) with water table within 1 m of the soil surface.
- **Haplosalids**—the hypersaline (salic horizon) soils (coastal & inland sabkha) without water table within 1 m of the soil surface.
- **Torriorthents**—the soils that have gravelly layers (gravels/sea shells > 35% by volume) within upper 1 m.
- **Torripsamments**—the soil that have deep desert sand without any accumulation.
- **Haplogypsids**—the soils that have gypsic horizon within upper 1 m.
- **Argigypsids**—the soils that have both the argillic and gypsic horizons within upper 1 m.
- **Gypsiargids**—the argids (with argillic horizon) that have also a gypsic horizon within 150 cm of the soil surface.

6.5 Soil Subgroups

At the fourth level of soil taxonomy hierarchy eighteen soil subgroups have been mapped in both reconnaissance and semi-detailed soil surveys of Kuwait (KISR 1999a, b). Typic Haplosalids was not recorded but likely to be recorded in future surveys work in Kuwait therefore, it is added in the list below and Table 6.2.

- **Typic Haplocalcids**—the soils with calcic horizon in the upper 1 m.
- **Typic Petrocalcids**—the soils with petrocalcic horizon in the upper 1 m.
- **Calcic Petrocalcids**—it is petrocalcids as defined above, but it has calcic horizon above the hardpan.
- **Typic Petrogypsids**—the soils that have gypsic hardpan (petrogypsic horizon) within upper 1 m.
- **Calcic Petrogypsids**—same as Petrogypsids defined above but has calcic horizon above the hardpan.
- **Petrocalic Petrogypsids**—same as Petrogypsids defined above but has petrocalcic horizon above the petrogypsic hardpan.
- **Petrocalcic Salidic Petrogypsids**—same as Petrocalcic Petrogypsids but has ECe of more than 8 to less than 30 dS m^{-1} in a layer 10 cm or more thick, within 100 cm of the soil surface.
- **Typic Calcigypsids**—the gypsids (with gypsic horizon) soil that also has a calcic horizon within 100 cm of the soil surface.
- **Lithic Calcigypsids**—Calcigypsids that have a lithic contact within 50 cm of the soil surface.
- **Leptic Haplogypsids**—a gypsic horizon starting in the upper 18 cm.
- **Typic Haplogypsids**—the soils that have gypsic horizon within upper 1 m but below 18 cm.
- **Calcic Salidic Argigypsids**—Argigypsids that have a calcic horizon overlying the gypsic horizon, but has ECe of more than 8 to less than 30 dS m^{-1} in a layer 10 cm or more thick, within 100 cm of the soil surface.
- **Typic Gypsiargids**—The argids (argillic horizon) that also have gypsic horizon within 150 cm of the soil surface.
- **Gypsic Aquisalids**—hypersaline soil (salic horizon) with gypsic horizon and water table in the upper 1 m.
- **Typic Aquisalids**—hypersaline soil (salic horizon) with water table in the upper 1 m.
- **Typic Haplosalids**—hypersaline soil (salic horizon) without water table in the upper 1 m.
- **Salidic Torriorthents**—sandy soil with high content of rock or shell fragments (exceeding 35%) in the upper 100 cm of the soil surface, and ECe of more than 8 to less than 30 dS m^{-1} in a layer 10 cm or more thick, within 100 cm of the soil surface.
- **Typic Torripsammnts**—the soil that have deep desert sand without any accumulation.
- **Lithic Torripsamments**—same as typic torripsamments but has lithic contact at 45–75 cm depth.

6.6 Soil Families

At the fifth level of soil taxonomy hierarchy, 22 soil families in the soil order Aridisols and only two in the soil order Entisols have been mapped in Kuwait. These are briefly described below.

6.6.1 Families in the Soil Order Aridisols

Gypsic Aquisalids, sandy, mixed, hyperthermic—hypersaline, sandy, gypsic, mixed mineralogy soil with water table within 1 m from soil surface.

Typic Aquisalids, sandy, mixed, hyperthermic—hypersaline, sandy, mixed mineralogy soil with water table within 1 m from soil surface.

Petrocalcic Petrogypsids, sandy, mixed, hyperthermic—Petrocalcic and petrogypsic hard pan within 1 m from soil surface, sandy soil with mixed mineralogy.

Petrocalcic Salidic Petrogypsids, sandy, mixed, hyperthermic (shallow)—Petrocalcic and petrogypsic hard pan within 25 to less than 50 cm from soil surface, sandy soil with mixed mineralogy and ECe of more than 8 to less than 30 dS m^{-1} in a layer 10 cm or more thick, within 100 cm of the soil surface.

Petrocalcic Salidic Petrogypsids, coarse-loamy, mixed, active, hyperthermic—Petrocalcic and petrogypsic hard pan within 1 m from soil surface, coarse-loamy soil with mixed mineralogy and ECe of more than 8 to less than 30 dS m^{-1} in a layer 10 cm or more thick, within 100 cm of the soil surface.

Calcic Salidic Petrogypsids, sandy, mixed, hyperthermic—calcic horizon above the petrogypsic horizon within upper 1 m from soil surface, sandy soil with mixed mineralogy and ECe of more than 8 to less than 30 dS m^{-1} in a layer 10 cm or more thick, within 100 cm of the soil surface.

Typic Petrogypsids, coarse-gypseous, hypergypsic, hyperthermic—petrogypsic horizon within upper 1 m from soil surface, coarse-gypseous soil with hypergypsic mineralogy.

Typic Petrogypsids, coarse-gypseous, hypergypsic, hyperthermic (shallow)—petrogypsic horizon within 25 to less than 50 cm from soil surface, coarse-gypseous soil with hypergypsic mineralogy.

Typic Calcigypsids, sandy, mixed, hyperthermic—both calcic and gypsic horizons within the upper 1 m from soil surface, sandy soil with mixed mineralogy.

Typic Calcigypsids, coarse-gypseous, hypergypsic, hyperthermic—both calcic and gypsic horizons within the upper 1 m from soil surface, coarse-gypseous sandy soil with hypergypsic mineralogy.

Typic Calcigypsids, coarse-loamy, mixed, semiactive, hyperthermic —both calcic and gypsic horizons within the upper 1 m from soil surface, coarse-loamy soil with mixed mineralogy.

Typic Haplocalcids, sandy, mixed, hyperthermic—calcic horizon within upper 1 m from soil surface, sandy soil with mixed mineralogy.

Typic Haplocalcids, coarse-loamy, mixed, semiactive, hyperthermic—calcic horizon within upper 1 m from soil surface, coarse-loamy soil with mixed mineralogy.

Leptic Haplogypsids, coarse-gypseous, hypergypsic, hyperthermic—gypsic horizon within 18 cm from soil surface, coarse-gypseous soil with hypergypsic mineralogy.

Typic Haplogypsids, sandy, mixed, hyperthermic—gypsic horizon between 18 and 100 cm from soil surface, sandy soil with mixed mineralogy.

Typic Haplogypsids, coarse-gypseous, hypergypsic, hyperthermic—gypsic horizon between 18 and 100 cm from soil surface, coarse-gypseous soil with gypsic mineralogy.

Typic Haplogypsids, coarse-loamy, mixed, semiactive, hyperthermic—gypsic horizon between 18 and 100 cm from soil surface, coarse loamy soil with mixed mineralogy.

Calcic Petrocalcids, sandy, mixed, hyperthermic—a calcic horizon above the petrocalcic horizon within 1 m from soil surface, sandy soil with mixed mineralogy.

Calcic Petrocalcids, sandy, mixed, hyperthermic (shallow)—calcic horizon above the petrocalcic horizon within 25 to less than 50 cm from soil surface, sandy soil with mixed mineralogy.

Calcic Petrocalcids, coarse-loamy, mixed, subactive, hyperthermic—a calcic horizon above the petrocalcic horizon within 1 m from soil surface, coarse loamy soil with mixed mineralogy.

Typic Petrocalcids, sandy, mixed, hyperthermic—a petrocalcic horizon within 1 m from soil surface, sandy soil with mixed mineralogy.

Typic Petrocalcids, sandy, mixed, hyperthermic (shallow)—a petrocalcic horizon within 25 to less than 50 cm from soil surface, sandy soil with mixed mineralogy.

6.6.2 Families in the Soil Order Entisols

Typic Torripsamments, mixed, hyperthermic—deep sandy soil, with mixed mineralogy.

Salidic Torriorthents, sandy-skeletal, carbonatic, hyperthermic—gravely (>35% by volume) sandy deep soil, with carbonatic mineralogy and ECe of more than 8 to less than 30 dS m^{-1} in a layer 10 cm or more thick, within 100 cm of the soil surface.

6.7 Important Definitions (Soil Survey Staff 2014) to Understand Soil Families Characteristics Recognized in Kuwait

6.7.1 Hypergypsic Mineralogy

The soils with horizons in the mineralogy control section that have a substitute class that replaces the particle-size class, other than fragmental, and that have 40 percent or more (by weight) gypsum either in the fine-earth fraction or in the fraction less than 20 mm in diameter, whichever has a higher percentage of gypsum.

6.7.2 Gypsic Mineralogy

Any particle-size class and 15 percent or more (by weight) gypsum, either in the fine-earth fraction (<2 mm soil) or in the fraction less than 20 mm in diameter, whichever has a higher percentage of gypsum.

6.7.3 Carbonatic Mineralogy

Any particle-size class and more than 40 percent (by weight) carbonates (expressed as $CaCO_3$) plus gypsum, either in the fine-earth fraction (< 2 mm soil) or in the fraction less than 20 mm in diameter, whichever has a higher percentage of carbonates plus gypsum.

6.7.4 Mixed Mineralogy

No specific dominance of minerals, but mixture of minerals presence in soil.

6.7.5 Shallow

The soil where root restricting layer occurs within 25 to less than 50 cm depth.

6.7.6 Coarse-Gypseous

Less than 35 percent (by volume) rock fragments and 50 percent or more (by weight) particles with diameters of 0.1 to 2.0 mm.

6.7.7 Sandy-Skeletal

Have a total content of rock fragments, plus any artifacts 2 mm or larger in diameter which are both cohesive and persistent, of 35 percent or more (by volume) *and* a texture class of coarse sand, sand, fine sand, loamy coarse sand, loamy sand, or loamy fine sand in the fine-earth fraction (<2 mm).

6.7.8 Sandy

Have a texture class of coarse sand, sand, fine sand, loamy coarse sand, loamy sand, or loamy fine sand in the fine-earth fraction.

6.7.9 Loamy

Have a texture class of loamy very fine sand, very fine sand, or finer, including less than 35 percent (by weight) clay in the fine-earth fraction (< 2 mm), and are in a shallow family or in a Lithic subgroup, or the layer is a part in a strongly contrasting particle-size class.

6.7.10 Coarse-Loamy

Have, in the fraction less than 75 mm in diameter, 15 percent or more (by weight) particles with diameters of 0.1 to 75 mm (fine sand or coarser, including gravel and artifacts 2 to 75 mm in diameter which are both cohesive and persistent) *and*, in the fine-earth fraction (<2 mm), less than 18 percent (by weight) clay.

6.7.11 Fine-Loamy

Have, in the fraction less than 75 mm in diameter, 15 percent or more (by weight) particles with diameters of 0.1 to 75 mm (fine sand or coarser, including gravel and artifacts 2 to 75 mm in diameter which are both cohesive and persistent) *and*, in the fine-earth fraction, 18 to less than 35 percent (by weight) clay.

6.7.12 Hyperthermic

Soils that have a difference in soil temperature of 6 °C or more between mean summer and mean winter and a mean annual soil temperature of 22 °C (72 °F) or higher.

References

KISR (1999a) Soil Survey for the State of Kuwait. Reconnaissance survey, vol II. Kuwait Institute for Scientific Research, Kuwait

KISR (1999b) Soil Survey for the State of Kuwait. Semi-detailed survey, vol IV. Kuwait Institute for Scientific Research, Kuwait

Soil Survey Staff (2014) Keys to soil taxonomy, 12th edn. US Department of Agriculture, Natural Resources Conservation Service, US Government Printing Office, Washington, DC

Moltkiopsis ciliate is perennial plant seen all over Kuwait, especially on loose sand "Entisols"

Laboratory Soil Procedures for Kuwait Soil Taxonomy

7

Abstract

Soil classification is based on field description of in-depth soil pits, augur holes and/or soil profiles supplemented with laboratory analyses of soil samples collected from different horizons and layers. The laboratory analysis includes physical, chemical, engineering and mineralogical to form the basis of the operational definitions for Soil Taxonomy as described in this chapter, and these methods have been used to support Kuwait Soil Taxonomy. Most of the methods are taken from Kellogg Soil Survey Laboratory Methods Manual, Soil Survey Investigations Report No. 42, Version 5.0, except where otherwise specified with each determination. In this chapter the procedures for various analyses (texture, coarse fragments, loss on acid treatment, calcium carbonate equivalents, gypsum, saturation extract analysis, sodium adsorption ratio (SAR), electrical conductivity of soil extract (ECe), pH of saturated soil paste (pHs), extractable cations, cation-exchange-capacity (CEC), exchangeable sodium percentage (ESP), water retention, bulk density, particle density, porosity, soil organic matter and organic carbon, engineering data-Atterberg limits (liquid limit, plastic limit, plasticity index), percent passing through different sieves, Unified Soil Classification System (USCS) and American Association of State Highways and Transportation Officials-AASHTO) have been described.

Keywords

Soil analysis · Physical · Chemical · Mineralogical · Engineering · Classification

S. A. Shahid and S. A. S. Omar, *Kuwait Soil Taxonomy*,
https://doi.org/10.1007/978-3-030-95297-6_7

7.1 Introduction

Kuwait Soil Taxonomy is supported by analyses of soil samples from georeferenced soil *PEDONS* identified in Kuwait. The pedon is the smallest body of one kind of soil large enough to represent the nature and arrangement of horizons and variability in the other properties that are preserved in samples (Soil Science Division Staff 2017). The soil procedure protocols of the United States Department of Agriculture-Natural Resources Conservation Service (USDA-NRCS) have been used. The soil analyses are mandatory to understand soil relationships and genesis in the landscapes to map soil heterogeneity in the form of map units as base to publish soil and thematic maps. The soil survey for the state of Kuwait was completed between 1995 and 1999 (KISR 1999a, b), when USDA-NRCS procedures outlined in the 1994 edition were used, which have been correlated in this book with the current "Kellogg Soil Survey Laboratory Methods Manual", Soil Survey Investigations Report No. 42 Version 5.0 (Soil Survey Staff 2014a). The soil procedures in this chapter are documented as reference to support future soil classification work in Kuwait. The physical, chemical, mineralogical and engineering analyses were carried out in the soil survey for the state of Kuwait, which can be accessed in the final reports (KISR 1999a, b), the laboratory procedures used are described in this chapter.

7.2 Soil Samples Collection, Preparation and Processing

The georeferenced soil samples collected from soil survey/soil classification work, from routine survey and the typical soil profiles/pits of identified soil taxa were transferred to the soil laboratories. A pit may be excavated by hand or with a backhoe for in-depth view of soil for description and collection of samples from different horizons to support soil survey and classification. The soil pits are generally excavated to a maximum depth of 2 m, if hard pan exists in the soil pit, then to upper depth of hard pan. The deeper soil pits (150 cm and over) are to be sloped upward to prevent entrapment and to avoid posing difficulty to come out of the soil pit. The plastic bags are recommended for sample collection, these are generally impermeable to CO_2 and O_2, preventing sample drying, i.e., aerobic samples will remain aerobic during transport to the laboratory. The sealable plastic bags are recommended to prevent loss of moisture, to allow determine field moisture content of soils. Each soil sample should be given a unique ID for reference.

Laboratory analyses of soil samples are generally performed on the air-dry, fine-earth (<2-mm) fraction, which is generally the optimum water content to handle and to process soil. Therefore, the soil samples are processed (drying, crushing, sieving) to obtain fine earth fraction (<2 mm) for physical, chemical, mineralogical and engineering analyses. This is due to the fact that the weight of air-dry soil remains relatively constant and the biological activity is low during storage.

Some standard air-dry analyses require non-sieved material, e.g., whole-soil samples for aggregate stability testing, and processing of < 2 mm soil sample for Atterberg Limits to 40-mesh (0.425 mm). It should be noted that the soil samples collected must represent soil taxa to be used in establishing soil map. Care must be exercised to avoid any contamination during sampling and processing, as well as it is recommended to use a mask, face-shield goggles, and/or respirators when operating dust generating equipment. The objective of soil samples preparation and processing is to homogenize and to obtain representative sample for analyses. Following are general equipment/apparatus required for sample processing.

- Electronic balance (±1 g)
- Samples drying trays
- Temperature controlled drier/oven (air-drying 30 to 35 °C for 3 to 7 days or less based on the soil moisture at collection)
- Set of square-hole sieves, with following openings
  - 2mm (10 mesh)
  - 4.75 mm (4 mesh)
  - 19 mm (3/4 inches)
  - 76 mm, 3 inches
- Soil samples storing containers

The units of measurement reported on the data sheets of Soil Survey of Kuwait (KISR 1999a, b) and to be used in the Kuwait Soil Taxonomy may not be SI (international system of units) units. The following are conversions to SI units of measurement:

$$
\begin{aligned}
1\ \text{meq/100 g} &= 1\ \text{cmol(+)/kg} \\
1\ \text{meq/liter} &= 1\ \text{mmol}(\pm)/\text{L} \\
1\ \text{mmho/cm} &= 1\ \text{dS m}^{-1} \\
15\ \text{bar} &= 1500\ \text{kPa} \\
1/3\ \text{bar} &= 33\ \text{kPa} \\
1/10\ \text{bar} &= 10\ \text{kPa} \\
1\ \text{percent} &= 10\ \text{g kg}^{-1}
\end{aligned}
$$

7.3 Particle Size Analysis—Soil Texture

Particle size analysis determines the proportion of various soil fractions. Soil texture is determined by measuring three primary soil particles (sand, silt, clay) in fine earth fraction (<2 mm). By plotting percent distribution of sand, silt and clay on textural triangle (Soil Science Division Staff 2017) soil textural class is determined (Fig. 7.1). Soil texture is used in soil taxonomy to report particle-size class for soil family and soil series. Soil texture is determined on less than 2 mm soil fraction, which receives pre-treatments to remove organic matter, mineral salts and gypsum. After giving the pre-treatments, the air-dried soil sample is dispersed with Na-hexametaphosphate solution and volume made in the cylinder. The fall of silt and clay fractions in settling suspension is recorded either by taking sample by Pipette or density measurement by hydrometer after certain times based on the Stoke's law. Sand is larger (>0.05–2 mm) in size and falls quickly in the suspension, thus cannot be determined directly by both Pipette and Hydrometer methods, but through difference:

$$\%\ \text{Sand} = 100 - (\%\ \text{silt} + \%\ \text{clay})$$

Results are presented as weight percentages of the fraction less than 2 mm on an oven-dry soil basis, including carbonates but excluding gypsum.

Sub-fractions of sand as defined by Soil Science Division Staff (2017) can be determined by wet sieving, to define following classes.

Very coarse sand = 1.0 to 2.0 mm
Coarse sand = 1.0 – 0.5 mm
Medium sand = 0.5 – 0.250 mm
Fine sand = 0.250 – 0.100 mm
Very fine sand = 0.100 – 0.05 mm

It is to be noted that exact sieve size may not be available, therefore near size can be used, such as, sieves 0.105, and 0.046 mm can be used instead of 0.100 and 0.05 mm respectively. Various sieves sizes in mm are also labeled as US No or Tyler mesh size, for example (1 mm = 18 US No & 16 mesh size; 0.5 mm (25 & 32), 0.25 mm (60 & 60), 0.105 (140 & 150), and 0.047 mm (300 & 300).

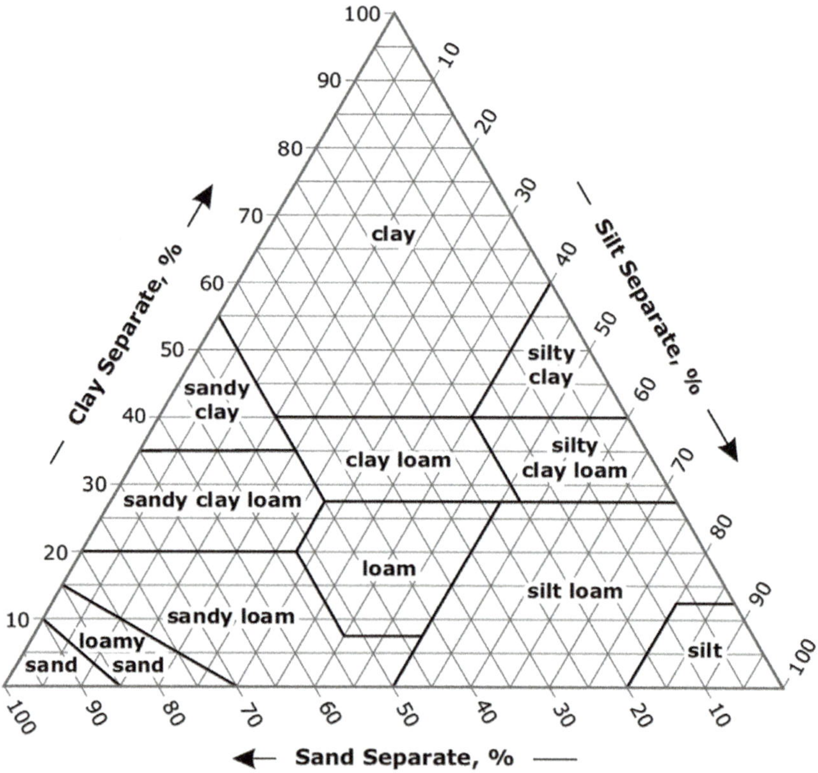

Fig. 7.1 USDA soil textural triangle (Schoeneberger et al. 2012)

7.4 Coarse Fragments

The material greater than 2 mm including rock fragments of geologic origin, carbonate nodules/concretions & shells, but excluding gypsum crystals, are sieved and weight measurements are made and recorded on the 20- to 75-mm, 5- to 20-mm, and 2- to 5-mm fractions, i.e., coarse, medium and fine gravels respectively. These coarse fragments are excluded from most chemical, physical, and mineralogical analyses, and are discarded, exception is for coarse fragments with carbonate- or gypsum-indurated material or material from Cr and R soil horizons. Such special composition coarse fractions are recombined with fine earth fraction, crushed, homogenized and stored for laboratory analysis.

7.5 Moisture Content

A known weight of air-dried soil is oven-dried (105 °C) overnight, and the difference in weight is reported as % moisture content. In case of gypsiferous soil, the moisture content must be adjusted by removing the crystal water of gypsum in soil (methods No. 3D1). Moisture content is required to convert laboratory analytical results from air-dry to oven-dry basis (a standard way of results presentation).

7.6 Loss on Acid Treatment (LAT)

The LAT is determined to validate the quantity of calcium carbonate equivalents in the soil as measured by either back titration or calcimeter method. A known weight of air-dried soil sample is reacted with 1 M HCl to decompose carbonates, the treated sample is oven-dried and reweighed to calculate percent loss in weight after treatment.

7.7 Calcium Carbonate ($CaCO_3$) Equivalents

The calcium carbonate ($CaCO_3$) equivalent (e.g., includes carbonates of Mg, Ca-Mg, Na, and K etc.) in soil is used in Kuwait Soil Taxonomy to identify calcic, petrocalcic horizons, and carbonatic mineralogy class (Soil Survey Staff 2014b). In addition, presence of calcium carbonate induces soil buffering capacity and provides adsorption and precipitation reactions for phosphorous and micronutrients. The calcium carbonate is reported on fine earth fraction (<2 mm) and the 2 to 20 mm basis. In the laboratory calcium carbonate equivalent is determined by two methods, (i) calcimeter and; (ii) back titration. In the calcimeter method, a known quantity of air-dried soil is reacted with HCl, the CO_2 produced through calcium carbonate decomposition is measured manometrically (calcimeter), which is used to calculate percent calcium carbonate on oven-dry soil basis (method no 4E1a1a1). In the back titration method, a known quantity of air-dry soil is reacted with known volume of 1 N HCl, the leftover HCl is back titrated with 1 N NaOH solution in the presence of phenolphthalein indicator. The volume of HCl consumed during reaction with soil is used to calculate % calcium carbonate equivalent in soil.

7.8 Gypsum ($CaSO_4.2H_2O$)

The quantity of gypsum ($CaSO_4.2H_2O$) present in the control section of soil profile is used to identify gypsic and petrogypsic horizons, and gypsum mineralogy class at the family level of US soil taxonomy hierarchy (Soil Survey Staff 2014b). Gypsum is reported on both a < 2 and a < 20-mm basis. To determine gypsum, a known weight of soil is mixed with water to dissolve gypsum. The soil suspension is filtered, and acetone is added to a portion of clear filtrate to precipitate dissolved gypsum. After centrifugation, the precipitated gypsum is dissolved in water and electrical conductivity of the gypsum solution is measured. The EC is used to calculate percent gypsum in soil.

7.9 Extractable Cations

The extractable cations (Ca^{2+}, Mg^{2+}, K^+, and Na^+) from the 1 N NH_4OAc (buffered-pH 7) extraction (method 4B1a1) are generally assumed to be sum of soluble and exchangeable cations. The extractable cations are measured by either Atomic Absorption Spectrophotometer (AAS) or Inductively Coupled Plasma (ICP) and expressed as meq/100 g (Thomas 1982) or SI unit [(cmol (+)/kg)]. In this method extraction of Ca^{2+} is overestimated due to Ca dissolved from free calcium carbonate and gypsum and K from dominance of illite/mica in soil (Thomas 1982).

7.10 Cation-Exchange-Capacity (CEC)

The cation-exchange-capacity is the ability of soil to hold and exchange cations in a plant available form (NH_4^+, K^+, Ca^{2+}, Mg^{2+} etc.). The CEC is also a measure of soil fertility level. The CEC is measured by two ways. *Method 1:* by saturating the exchange sites of soil with an index cation (NH_4), washing the excess NH_4 with 95% ethanol, and displacement of adsorbed index NH_4 by 2M KCl, analyzed by steam distillation and titration to determine the NH_4^+ adsorbed on the soil exchange complex and reported as meq/100g or cmol(+)/kg (method no. 4B1a1a1a1). *Method 2:* In this method soil exchange complex is equilibrated with 1 N NaOAC (buffered at pH 8.2), the excess NaOAC is washed with 95% ethanol, finally the Na is extracted with 1 N NH_4OAC (pH 7) and Na is measured using any of the three equipment (flame photometer, atomic absorption spectrophotometer, inductively coupled plasma) and CEC reported as meq/100 g or cmol (+)/kg.

7.11 Exchangeable Sodium Percentage (ESP)

The ESP is used to identify soil sodicity problem, if ESP > 15, the soil is classified as sodic or alkali soil. In soil taxonomy Sodium Adsorption Ratio is used as criteria for natric horizon (Soil Survey Staff 2014b). The exchangeable sodium percentage (ESP) can be calculated by two calculation procedures.

Method 1: Compute the exchangeable sodium percentage (ESP) by dividing the exchangeable sodium (ES) by the CEC measured (meq/100 g) by NH_4OAc, pH 7.0 (CEC–7) and multiplying by 100 (method no 4F3a1).

$$\mathrm{ESP} = [\mathrm{Exchangeable\ Na/CEC}] \times 100.$$

Method 2: Compute the sodium adsorption ratio (SAR) value in the following formula given by U. S. Salinity Laboratory Staff (1954).

$$\mathrm{ESP} = [100 \times (-0.0126 + 0.01475 \times \mathrm{SAR})/(1 + (-0.0126 + 0.01475 \times \mathrm{SAR})].$$

The SAR is calculated by using Na, Ca, Mg (meq/l) from extract of saturated soil paste.

$$SAR = Na/[(Ca + Mg)/2)]^{0.5}$$

Where Na, Ca + Mg are reported as meq/l and the SAR $(mmoles/l)^{0.5}$

7.12 Saturation Percentage (SP)

The saturation percentage indicates available water content in soil. In addition, the SP is used to convert soil solution chemistry data (from saturation extract) to soil weight basis (e.g., meq/l to meq/100 g). A 250–300 g air-dried soil is used to prepare a standard saturated soil paste. A portion (30–50 g) of saturated soil paste is used to determine the moisture content (oven-drying at 110 °C for 12 to 16 h) to determine saturation percentage (SP) by method 4F2a1. The SP can also be calculated by using the volume of water to prepare saturated soil paste.

7.13 Preparation of Saturated Soil Paste

A portion of 250 g air-dry soil is used to prepare saturated soil paste by adding deionized water and mixing, to reach to a level where soil paste glistens as it reflects light, flows slightly when the container is tipped, and slides freely and cleanly off the spatula. The saturated paste is covered and allowed to stand overnight. The saturation criteria are then rechecked.

7.14 Saturation Extract Analysis

The extract collected from saturated soil paste under vacuum (Fig. 7.2) is used to measure soluble cations (Na^+, K^+, Ca^{2+}, Mg^{2+}) and soluble anions (CO_3^{2-}, HCO_3^-, Cl^-, SO_4^{2-}). The soluble Na^+, K^+, Ca^{2+}, Mg^{2+}can be measured by AAS-ICP. Soluble HCO_3^- by titration with 0.011 N H_2SO_4 in the presence of methyl orange indicator, CO_3^{2-} by titration with 0.011 N H_2SO_4 in the presence of phenolphthalein indicator, and soluble Cl^-, NO_3^-, PO_4^{3-}, and SO_4^{2-} by ion chromatography. The Cl^- can also be measured by chloride analyzer.

7.15 Electrical Conductivity of Soil Saturation Extract (ECe)

The electrical conductivity of the saturation extract (ECe) is used as a criterion for classifying a soil as saline (ECe > 4 dS m^{-1}). Other uses of ECe measurement include the estimation of the total cation concentration in the extract, salt percentage in solution (Psw), salt percentage in soil (Pss), and osmotic pressure (OP). The ECe is measured using a pre-calibrated EC meter, and reported as deci Siemens per meter (dS m^{-1}) or milli Siemens per centimeter (mS cm^{-1}), both are equal (1 dS m^{-1} = 1 mS cm^{-1}).

7.16 Osmotic Potential (OP)

Osmotic potential (atmospheres) provides a guide to the energy a plant must expend to extract water from soil solution; a more saline solution requires more energy. The ECe value was used to estimate the osmotic potential in atmospheres

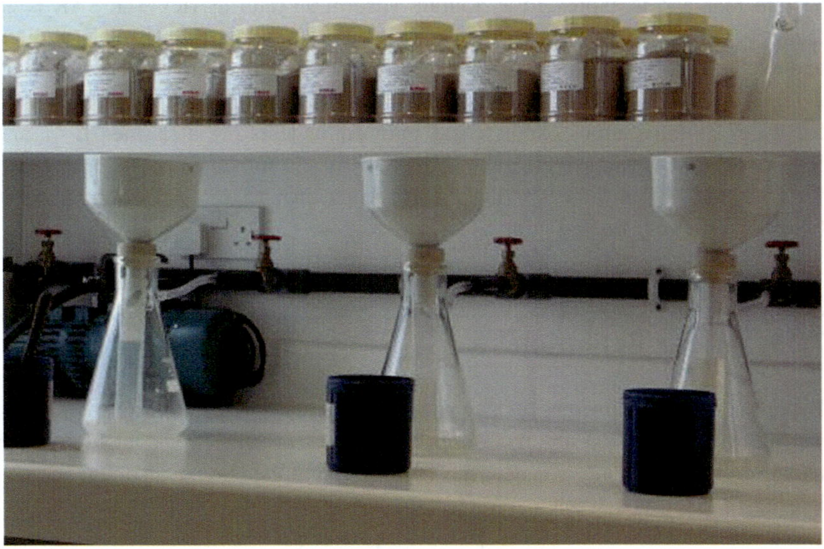

Fig. 7.2 Vacuum extraction of soil extract from saturated soil paste

of a solution as OP $\approx$ 0.36 $\times$ ECe where ECe is expressed as dS m^{-1} (U.S. Salinity Laboratory Staff 1954).

7.17 Soil Reaction or Hydrogen Ion Activity (pH)

The soil reaction is an expression of the degree of acidity or alkalinity of a soil and is expressed as pH. The 1:1 water pH (4C1a2a1) is commonly performed soil pH measurement in the field and the laboratory. A 20-g soil sample is mixed with 20 mL of deionized water (1:1 w:v) with occasional stirring. The pH of saturated soil paste (pHs) is measured with a standard pH meter calibrated using buffer solutions of pH 4.0, 7.0, and 10.0. These pH values are used as criteria for reaction classes in some taxonomic families (Soil Science Division Staff 2017; Soil Survey Staff 2014b; Schoeneberger et al. 2012).

7.18 Sodium Adsorption Ratio (SAR)

Sodicity is a measure of sodium ions in soil water, relative to calcium and magnesium ions and is expressed as sodium adsorption ratio (SAR). The SAR is an indirect measure of the equilibrium relation between sodium in the salt solution and exchangeable sodium adsorbed on the soil exchange complex. The SAR can be calculated by standard formula: SAR = $Na/[(Ca + Mg)/2]^{0.5}$, using inputs for the water soluble cations expressed as milli equivalents per liter (meq L^{-1}), and SAR expressed as (mmoles $L^{-1})^{0.5}$. If the SAR of the soil equals or is greater than 13 $(\text{mmoles/L})^{0.5}$, or the ESP equals or is greater than 15, the soil is termed sodic (USSL Staff 1954; Shahid et al. 2014).

7.19 Water Retention

Water retention is defined as the soil water content at a given soil water suction. By varying the soil suction and recording the changes in soil water content, a water retention function or curve is determined. A pressure membrane apparatus method is used to determine water retention (percent) at 1/10, 1/3, and 15 bars, respectively, for sieved, < 2 mm, air-dry soil samples (Fig. 7.3).

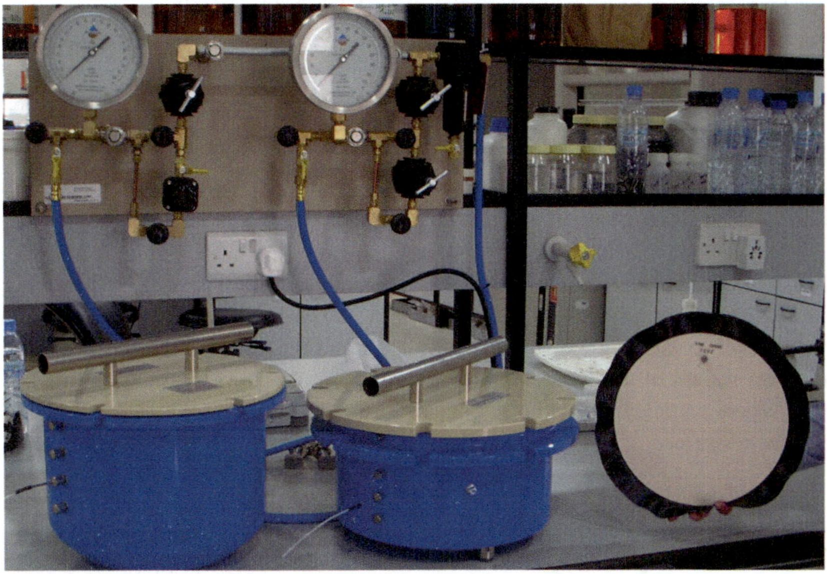

Fig. 7.3 Pressure membrane apparatus. Porous ceramic plate is also evident

Air-dry <2-mm (sieved) soil sample is placed in a retainer ring placed on a porous ceramic plate in a pressure-plate extractor. The plate is covered with water to wet the samples by capillarity. The sample is equilibrated at the specified pressures. The pressure is kept constant until equilibrium is obtained (Klute 1986). The gravimetric water content is determined.

The following are SI conversions:

$$\begin{aligned} 15\,\text{bar} &= 1,500\ \text{kPa} \\ 1/3\ \text{bar} &= 33\ \text{kPa} \\ 1/10\ \text{bar} &= 10\ \text{kPa} \end{aligned}$$

7.20 Bulk Density (BD)

Bulk density is the weight of soil (oven dry) per unit bulk volume of soil (volume occupied by soil solids and pore spaces). The bulk density of a soil indicates the pore space available for water and roots. For example, a bulk density of 1.6 g cm^{-3}

and higher can restrict water storage and root penetration. Bulk density samples can be collected in standard steel cores of known volume and then oven dried. The loss in weight on drying is recorded and bulk density calculated by standard formula.

7.21 Particle Density (PD)

Density is defined as mass per unit volume. Particle density refers to the density of the solid particles collectively (Flint and Flint 2002). The method requires using a pycnometer to estimate the volume of water or an inert gas (e.g., He) displaced by a known weight of soil, ensuring that all air is removed from the mixture (method 3G). Where particle density has not been measured, a default value of 2.65 g cm^{-3} has been assumed as the basis of other calculations.

7.22 Porosity

The percent volume of bulk soil that is occupied by pore spaces (the space filled by air or water) is called porosity. Porosity is calculated by using a standard formula of 100 (1–BD/PD), where BD and PD represent the bulk density and particle density, respectively.

7.23 Soil Organic Matter and Organic Carbon

The mineral content consists of the plant ash and soil particles that remain after removal of organic matter. Total soil carbon is the cumulative carbon from both sources, the organic and inorganic carbon (C). Most of the organic C is associated with the organic matter fraction, and the inorganic C is generally found with carbonate minerals. The soil organic matter generally includes only those organic materials that accompany soil particles through a 2-mm sieve (Nelson and Sommers 1982). Organic C is a major component of soil organic matter, which is determined generally by following methods.

Method 1: Loss on ignition (LOI) method. The soil material containing organic matter is ignited at 400 °C, and loss in weight is considered as organic matter (5A). The percent organic matter is calculated by difference (i.e., 100—percent mineral content). This quantity can be used in place of organic matter estimates by the Walkley–Black Organic C method.

Method 2: Organic carbon is mostly determined by wet digestion (Walkley 1947) method (6A1). The organic carbon is then multiplied by the Van Bemmelen factor of 1.724 to estimate percent organic matter based upon the assumption that soil organic matter contains 58% organic C.

7.24 Engineering Data

The soils can be analyzed for their Atterberg limits and classified for engineering purposes according to the ASTM Unified Soil Classification System (USCS) (Gee and Bauder 1986), and the system adopted by the American Association of State Highway and Transportation Officials (AASHTO).

7.24.1 Atterberg Limits

The liquid limit and plastic limits are called Atterberg limits. The < 2-mm soil is further processed (<0.425 mm for Atterberg limits and ≈180 μm) so as to be in accordance with a standard method, e.g., Atterberg limits.

7.24.1.1 Liquid Limit (LL)

The liquid limit is the water content when soil starts to behave as a liquid. It is measured by placing soil sample in a standard cup and making a separation (groove) using a grooving tool. The cup is dropped till the separation vanishes. The water content of the soil is obtained from this sample to measure liquid limit. The LL is reported as percent water on a < 0.425-mm basis (40-mesh) (Fig. 7.4).

7.24.1.2 Plastic Limit (PL)

The plastic limit is the moisture content where the soil thread breaks apart at a diameter of 3.2 mm. A soil is considered non-plastic if a thread cannot be rolled out down to 3.2 mm at any moisture possible (Fig. 7.5).

7.24.1.3 Plasticity Index (PI)

Numerically, the Plasticity Index (PI) is the difference in the water content between the LL and the plastic limit (PL). The PI is reported as percent water on a < 0.425-mm basis.

Fig. 7.4 Liquid limit measurement device and grooved soil sample

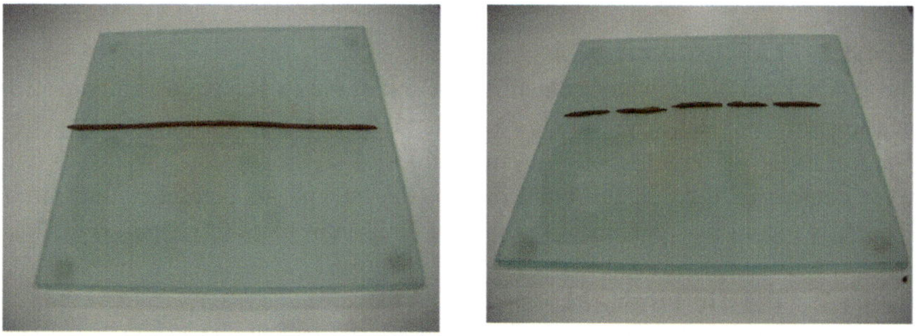

A rolled thread of soil Soil thread breaks apart

Fig. 7.5 Plastic limit measurement process

7.24.2 Percent Passing Sieves

An air-dried soil subsample can be sieved through a nest of sieves, mesh numbers 4 (4.75 mm), 10 (2 mm), 40 (0.42 mm), and 200 (0.074 mm), and shaken mechanically. The material retained on each sieve can be weighed, and calculations can be made using a standard formula (USDA-NRCS 1995) to represent percent soil passing the respective sieves.

7.24.3 Unified Soil Classification System (USCS)

The USCS is a system for classifying mineral and organic soils for engineering purposes based on particle-size characteristics, the amounts of various sizes, and the characteristics of the very fine particles assessed by the liquid limit and plasticity index.

The USCS identifies three major soil divisions:

(i) coarse-grained soils having less than 50 percent, by weight, of particles smaller than 0.074 mm in diameter;
(ii) fine-grained soils having 50 percent or more, by weight, of particles smaller than 0.074 mm in diameter; and
(iii) highly organic soils that demonstrate certain organic characteristics.

These divisions are further subdivided into 15 basic soil groups. The major soil divisions and basic soil groups are determined on the basis of estimated or measured values for grain-size distribution and Atterberg limits.

The various groupings of the USCS classification have been devised to correlate with the engineering behavior of soils. This correlation provides a useful first step in any field or laboratory investigation of soils for engineering purposes. The classification and its interpretation are applicable in the design of foundations for a range of structures. The dominant USCS classifications for Kuwait soils and their inherent characteristics are:

SM: Silty sand with a maximum of 30% fine material that is non-plastic. These soils contain various cementing agents, such as calcium carbonate, gypsum, clay, and silt. The degree of cementation can range from non-cemented to strongly cemented in both the horizontal and vertical directions (note that within the soil profile various horizons of SM material can exhibit differing degrees of cementation). From an engineering point of view, the ***solum*** would classify as SM. The strongly cemented material is often called "gatch" in Kuwait. The soil performs well upon remoulding and compacting. Gatch is widely used for engineering purpose.

SP-SM: Silty sand with fine material limited to between 5 and 12%. These soils contain various cementing agents, such as calcium carbonate, gypsum, clay, and silt. The soils are relatively permeable due to poor gradation. These soils are suitable for general engineering purposes.

SP: Poorly graded or gap graded fine to medium sand with little or no fines; not recommended for engineering purposes; difficult to compact; suitable for highly permeable fill material.

SW: Well graded siliceous sand with little or no fine material; very good material for construction and general engineering purposes.

CL: Inorganic clay. Typically mixed with sands with low plasticity and Wl < 50%. The soils are not common in Kuwait. They are found in localized areas in the north of the country on Bubiyan Island and around the coastal margins. In Kuwait this material is considered unsuitable for general engineering purposes.

MH: Inorganic silts with high plasticity. Very rare in Kuwait. In Kuwait this material is considered unsuitable for general engineering purposes.

CH: Inorganic clays with high plasticity. Very rare in Kuwait. In Kuwait this material is considered unsuitable for general engineering purposes.

7.24.4 AASHTO Group Classification

The AASHTO group classification is used for geotechnical engineering purposes such as highways and airfield construction and maintenance. It can be used to make general interpretations relating to the engineering performance of soil for these uses. The system is based on particle-size distribution and Atterberg limits and consists of a symbol and a group index. The classification is based on the <75 mm portion of the soil. The system identifies two general groups:

- Granular materials having 35 percent or less, by weight, particles smaller than 0.074 mm in diameter; and
- Silty-clay materials having more than 35 percent, by weight, particles smaller than 0.074 mm in diameter.

These two divisions are further subdivided into seven main group classifications. The group and subgroup classifications are based on estimated or measured grain-size distribution and on liquid limit and plasticity index values derived from measured values of samples taken from typical profiles.

The AASHTO group and subgroup classifications may be further modified by the addition of a group index value. The empirical group index formula was devised for approximate within-group evaluation of "clayey granular materials" and "silty-clay" materials.

The dominant AASHTO group classifications found in Kuwait and their inherent characteristics are:

A-2-4: Silty sand with a maximum silt content of 35 percent. The material is either non-plastic or has a low plasticity. The high sand content allows the soils to be classified as structural fill material and hence used for general engineering purposes (herein defined as construction and earthworks purposes).

A-1-b: This soil contains stone fragments and up to 25 percent fines (very fine sand, silt, and clay). They are either non-plastic or slightly plastic. The soil is suitable as general backfill for engineering purposes.

A-1-a: These soils are similar to those classified as A-1-b but are limited by the high percentage of stone fragments.

A3: Silty sand with a maximum of 10 percent silt. Usually non-plastic. This soil is suitable for structural and general backfill material.

A4: Silty soil with more than 35 percent fines. Moderately plastic. Generally, found near shore lines and within saline sabkha flats. This material is considered unsuitable for general engineering purposes in Kuwait.

A6: Clayey material with more than 35 percent fines. This material is highly plastic and is not suitable for general engineering purposes. Its extent is very limited in Kuwait.

A7: Similar to A-6 with high plasticity index. Includes expansive clays that are very rare in Kuwait. The material is considered not suitable for general engineering purposes in Kuwait.

7.25 Soil Mineralogy

X-ray diffraction analysis (XRDA), differential thermal analysis (DTA) and infrared analysis (IRA) are the mostly used methods to analyze soil and clay samples for mineral assemblages. However, XRD is considered the most powerful technique to identify mineral species in soils. Therefore, XRD is to be adopted to establish whole soil and clay mineralogy of the soil samples collected from the control sections of the identified soil families and soil series. For whole soil mineralogy, the soil samples are ground in a pestle and mortar and filled into an aluminum holder and pressed for preferred orientation of the soil minerals to scan through x-ray beam in the presence of Cu-Kα radiation to record x-ray patterns through three slits (variable divergent, receiving and scattering). The XRD scan of the sample was done stepwise using 1°2θ/min speed (Burt 2004). A Philips x-ray diffraction model PW/1800 microprocessor, with Ni filter, Cu-Kα radiation ($\lambda = 0.154$ nm) at 40 kV, 55 mA, was used. Diffraction peaks between 2 and 60° 2θ were recorded. The corresponding d-spacing and the relative intensities (I/Io) were calculated and compared with standard patterns (ASTM cards) using finks index (Finks Index). The XRD peaks were interpreted on semi-quantitative basis (D-Dominant > 60%, Co-dominant-sum of components > 60%, SD-sub-dominant (20–60%), M-minor (5–20%), T-traces < 5% & nd-not detected.

7.26 Clay Mineralogy

Clay (particles <2 μm) is a reactive (high surface area) soil component that controls physical, chemical, and nutritional behavior (nutrient storage) of soil and therefore affects agricultural potential. The clay-size material was separated from soil after the cementing material (organic matter, gypsum, and soluble salts) was removed for dispersion of clay in a soil suspension. Sedimentation and sampling for clay was based on Stoke's Law as described by Shahid and Jenkins (1991).

To resolve the clay mineral species, the clay was treated with K, K-heating at 300 and 550 °C, Mg, Mg ethylene glycol, and Mg-glycerol solvation. The treated clay samples were mounted on glass slides, and x-ray diffraction patterns were recorded using XRD equipment as above. The XRD patterns were interpreted for mineral species, comparing with standard reference patterns and measurements (ASTM cards). An additional test was made to confirm the presence of kaolinite using the Lim test (Lim et al. 1980). In this test the 7.2 Å peak of first order kaolinite shifts to 11.2 Å; however, second order chlorite peak remained at 7.00 Å. Based on the peak heights of the minerals, area under the peaks, and sharpness, the amounts of each mineral was estimated semi-quantitatively (%). The procedure and interpretation of XRD peaks for various clay minerals are very well described by Shahid (1989).

For a more complete list of d-spacings for confirmation or identification of a mineral, consult the Mineral Powder Diffraction File–Data Book (JCPDS 1980).

Following X-Ray criteria can be used to identify major clay and other minerals species in the soils of Kuwait.

7.26.1 X-Ray Diffraction Criteria

Kaolinite

a. Crystal structure missing at 500 °C
b. 7 Å (7.2 to 7.5 Å) with all other treatments
c. Is there a 7 Å peak? Is it destroyed at 500 °C? If so, the mineral is kaolinite
d. Is the peak sharp and at ≈7.1 Å (but absent at 500 °C)? If so, the mineral is kaolinite.
e. Is the peak at ≈7.1 Å with all treatments shifts completely to 11.2Å with Lim test, the mineral is kaolinite.

Mica (Illite)

a. 10 Å with all treatments
b. 10 Å with Mg^{2+}-saturation
c. Is there a 10 Å peak with Mg^{2+}-saturation? If so, the mineral is mica (illite).

Chlorite

a. Crystal structure of Fe-chlorites destroyed at 650 to 700 °C
b. 14 Å with all other treatments
c. 14 Å at 500 °C
d. Generally, also has a strong 7 Å peak
e. Is there a 14 Å peak when heated to 500 °C? If so, the mineral is chlorite.

Smectite

a. 14 Å with Mg^{2+}-saturation
b. 12 to 12.5 Å with K^+ or Na^+ saturation
c. 17 to 18 Å with Mg^{2+}glycerol solvation
d. 10 Å with K^+ saturation and heating to 300 °C
e. Is there a 17 to 18 Å peak upon solvation? If so, the mineral is smectite.

Palygorskite

a. 10.5 Å with all treatment
b. 10.5 Å peak shifted to 10 Å when heated 550 °C for 1 h
c. If, so, mineral is palygorskite

Quartz

a. Peaks at 4.27 and 3.34 Å with all treatments (only 3.34 Å if small amounts)

Potassium Feldspar

a. Peak at 3.24 Å with all treatments

Plagioclase Feldspar

a. Twin peaks between 3.16 and 3.21 Å with all treatments

Calcite

a. Peak at 3.035 Å with all treatments

Dolomite

a. Peak at 2.88 to 2.89 Å with all treatments

Gypsum

a. Peak at 7.56 Å with Mg^{2+} and Mg^{2+}-glycerol but destroyed when heated to 300 °C

Standard tables to convert θ or 2θ angles to crystal d-spacings are published in the U.S. Geological Survey Circular 29 (Switzer et al. 1948) and in other publications (Brown 1980).

7.27 X-Ray Fluorescence (XRF)

X-ray emission spectrography is elemental analysis by measuring the X-ray fluorescence produced by bombarding a sample with high-energy X-rays (Vanden 1965). To analyze soil sample, 1 g soil sample was ignited at 1,100 °C and dissolved in 9 g of lithium tetraborate and lithium borate (66:34) and 0.5 g of lanthanum oxide flux to form a disc for x-ray fluorescence analysis. The information is presented as percentages of elements and elemental oxide.

References

Brown, G. 1980. Appendix I (Tables for the determination of d in Å from 2θ for the Kα and Kβ radiations of copper, cobalt, and iron). In: Brindley GW, Brown G (eds) Crystal structures of clay minerals and their x-ray identification, pp 439–475. Mineralogical Soc. Monogr No. 5. Mineralogical Soc, Great Britain

Burt R (2004) Soil survey laboratory methods manual. Soil Survey Investigations Report No. 42, Version 4.0. U.S. Department of Agriculture, Natural Resources Conservation Service, National Soil Survey Center, Nebraska

Flint AL, Flint LE (2002) Particle density. In: Dane JH, Topp GC (eds) Methods of soil analysis, part 4: physical methods, pp 229–240. Soil Sci. Am. Book Series No. 5. ASA and SSSA, Madison, WI

Gee GW, Bauder JW (1986) Particle-size analysis. In: chapter 15: methods of soil analysis, part I: physical and mineralogical methods, 2nd edn, pp 383–411. Agronomy Monograph No. 9

JCPDS (1980) Mineral powder diffraction file–data book. International Centre for Diffraction Data, Swarthmore, PA

KISR (1999a) Soil survey for the state of Kuwait: reconnaissance survey, vol II. Kuwait Institute for Scientific Research, Kuwait

KISR (1999b) Soil survey for the state of Kuwait: semi-detailed survey, vol IV. Kuwait Institute for Scientific Research, Kuwait

Klute A (1986) Water retention: laboratory methods. In: Klute A (ed) Methods of soil analysis, part 1 physical and mineralogical methods. Agron. Monogr. 9. ASA and SSSA, Madison, WI

Lim CH, Jackson ML, Koons RD, Helmke PA (1980) Kaolins: sources of differences in cation exchange capacities and cesium retention. Clays Clay Miner 28:223–229

Nelson DW, Sommers LE (1982) Total carbon, organic carbon, and organic matter. In: Page AL, Miller RH, Keeney DR (eds) Methods of soil analysis, part 2 chemical and microbiological properties, 2nd edn, pp 539–579. Agronomy 9

Schoeneberger PJ, Wysocki DA, Benham EC, Soil Survey Staff (2012) Field book for describing and sampling soils, Version 3.0. USDA–NRCS, National Soil Survey Center, Lincoln, NE

Shahid SA (1989) Clay, silt and sand mineralogical investigations of soil profiles. In: Chapter 5, Mineralogy and micromorphology of salt-affected soils in Pakistan, vol 2, PhD Thesis. University of Bangor Wales, UK

Shahid SA, Jenkins DA (1991) Turbidometric technique for the measurement of various subdivisions of clay in soil. Pak J Agric Sci 28(1):91–98

Shahid SA, Zaman M, Heng L (2014) Introduction to soil salinity, sodicity and diagnostics techniques. In: Zaman et al (eds) Chapter 1, guideline for salinity assessment, mitigation and adaptation using nuclear and related techniques. https://doi.org/10.1007/978-3-319-96190-3_1

Soil Science Division Staff (2017) Soil survey manual. United States Department of Agriculture Handbook No. 18

Soil Survey Staff (2014a) Kellogg soil survey laboratory methods manual. Soil Survey Investigations Report No. 42, Version 5.0. R. Burt and Soil Survey Staff (ed.). U.S. Department of Agriculture, Natural Resources Conservation Service.

Soil Survey Staff (2014b) Keys to soil taxonomy, 12th edn. US Department of Agriculture, Natural Resources Conservation Service, US Government Printing Office, Washington DC

Switzer G, Axelrod JM, Lindberg ML, Larsen ES (1948) US Dept. Interior. Geological Survey. Circular 29. Washington, DC

Thomas GW (1982). Exchangeable cations. In: Page AL et al (ed) Methods of soil analysis, part 2 Chemical and microbiological properties, 2nd edn, pp 159–166. Agronomy monograph No. 9. Madison. American Society of Agronomy, Inc., Soil Science Society of America, Inc.

USDA-NRCS (1995) Soil Survey Laboratory Information Manual. Soil Survey Investigation Report No. 45. Version 1.0. United States Department of Agriculture, Natural Resources Conservation Service, National Soil Survey Center

US Salinity Laboratory Staff (1954) In: Richards LA (ed) Diagnosis and improvement of saline and alkali soils, 160p. U.S. Dept. of Agric. Handb. 60. U.S. Govt. Print Office, Washington, DC

Vanden HRC (1965) Elemental analysis by X−ray emission spectrography. In: Methods of soil analysis, vol 2, pp 771–821. Monograph No. 9. Amer. Soc. Agron. Madison, Wis

Walkley A (1947) A critical examination of a rapid method for determining organic carbon in soils: effect of variations in digestion conditions and of inorganic soil constituents. Soil Sci 63:251–263

Gynandriris sisyrinchium is perennial herb with bulb (rhizome). It is grown in sandy desert soils which are protected from external influence

Glossary of Terms Used in Soil Survey and Soil Classification

In this glossary, terms are included which are either used in soil survey and soil classification of Kuwait or in Kuwait Soil Taxonomy, and have relevance to future soil surveys in Kuwait. Some definitions are slightly modified/briefed to fit Kuwait soil conditions. The terms and definitions are taken from Ditzler (2017).

Abrupt textural change—A diagnostic soil characteristic of mineral soils defined as a considerable increase in silicate clay content within a short vertical distance (<7.5 cm) between the epipedon and an underlying argillic horizon. The required clay increase varies depending on the clay content of the epipedon. Where present, this feature significantly restricts the downward movement of water.

Accumulation—A term used in Soil Taxonomy and in the definitions of horizon suffix symbols to indicate that a specific material is present in a higher concentration than assumed for the original parent material. For example, a calcic horizon (Bk) has more calcium carbonate than is presumed to have been present in the parent material.

Apparent cation-exchange capacity (ACEC)—The cation-exchange capacity expressed as centimoles of charge per kilogram of clay; calculated as [(CEC-7/clay %) × 100]. It differs from *cation-exchange capacity:* CEC is measured for the *fine-earth fraction* and reported on a kg of soil basis while ACEC is a mathematical adjustment of CEC performed to report it on a kg of clay basis.

S. A. Shahid and S. A. S. Omar, *Kuwait Soil Taxonomy*,
https://doi.org/10.1007/978-3-030-95297-6

Apparent effective cation-exchange capacity (AECEC)—The effective cation-exchange capacity expressed as centimoles of charge per kilogram of clay; calculated as [(ECEC/clay %) × 100]. It differs from ECEC: ECEC is measured for the *fine-earth fraction* and reported on a kg of soil basis while AECEC is a mathematical adjustment of ECEC performed to report it on a kg of clay basis.

Apparent field texture—A tactile evaluation of soil texture in the field, with no inference as to expected laboratory results for textural analysis. Some soil samples cannot be adequately dispersed in the laboratory for accurate texture analysis using standard procedures. Factors that hinder the effectiveness of standard procedures for particle-size analysis include high content of low-activity clays or soils having carbonates, and gypsiferous materials. These soils tend to have ratios of 1500 kPa water to clay outside the normally expected range of 0.25 to 0.6, and clay content is therefore estimated rather than measured for family particle-size class as indicated in Soil Taxonomy. The apparent field texture reported in the soil profile description commonly appears coarser than the calculated result for placement in a taxonomic family particle-size class (e.g., fine sandy loam apparent field texture vs. fine family particle-size class).

Aquic conditions—A diagnostic soil characteristic of mineral soils that indicates the presence of continuous or periodic saturation and reduction in one or more layers, which leads to biogeochemical reduction of Mn and Fe. The criteria include saturation, reduction, and (for mineral soils) the presence of *redoximorphic features*.

Argillan—A ped or void surface feature composed of illuvial clay. The term is used interchangeably with *clay film*.

Association (map unit)—A map unit consisting of two or more dissimilar major components occurring in a regular and repeating pattern on the landscape. The major components of an association can be separated at a scale of about 1:24,000, but due to land use, or user needs, the map unit design integrates the predictable and repeating pattern of soil occurrence. Many general soil maps use soil associations as map units because they are at scales much smaller than 1:24,000 and only the characteristic landscapes of associated soils can be depicted, not the individual soils. The major components are sufficiently different in morphology or behavior that the map unit cannot be called a *consociation*.

Available water capacity—The volume of water that would be expected to be available to plants if the moisture content of the soil profile is at field capacity. It is generally considered to be the amount of water held between field capacity and the permanent wilting point (i.e., water retention difference) with corrections for factors such as salinity, rock fragments, and rooting depth. It is expressed as a volume fraction or as thickness of available water in a 100-cm or 150-cm soil profile. Physical and chemical factors of the soil that affect the ability of plants to extract water include texture, rock fragment content, bulk density, porosity, salt content, clay mineral types, organic matter content, low available calcium, high extractable aluminum, and presence of root-limiting layers (such as fragipans, duripans, and bedrock). Individual plant species also have somewhat differing abilities to extract water from the soil.

Bulk density—Mass per unit volume (commonly reported as g cm^{-3}) at a specified moisture content (10 kPa, 33 kPa, or oven dry). The sample measured includes the solids plus the pore space. Bulk density is used in the definitions of mineral soils, and several taxonomic classes. It is also used in estimating Ksat and is essential for converting from a weight to a volume basis, such as for estimating carbon content per cubic meter. Accurate bulk density determination requires careful sampling and transport of samples to preserve their natural volume.

Buried horizon—A genetic soil horizon that originally formed at or near the soil surface and is now covered by more recently deposited material. The major genetic features of the horizon were formed before burial. A buried horizon within a profile description is denoted with horizon symbol suffix "b" (e.g., Ab, Btb). The presence of one or more buried horizons does not automatically mean a buried soil is present for classification purposes. For a soil to be considered a buried soil, the technical criteria given in Chap. 4 of the *Keys to Soil Taxonomy* must be met. See *buried soil.*

Buried soil—A soil profile having one or more genetic horizons that is covered with a surface mantle of new soil material 50 cm or more thick. In essence, two soils are present—a new soil in the mantle and an older soil below that formed before the mantle was deposited.

Calcareous class—A soil family class term used in a limited number of mineral soil taxa that require a calcareous and reaction class term. Soils placed in the calcareous class effervesce in cold, dilute (0.1 M) HCl in the fine-earth fraction of all parts of the control section, as defined for calcareous and reaction classes. See *calcareous soil.*

Calcareous soil—A general term for a soil containing sufficient calcium carbonate to effervesce noticeably when treated with cold, dilute (0.1 M) HCl. There is no minimum amount of calcium carbonate required to be considered calcareous. The presence of calcareous horizons in the upper part of the soil is used as a criterion in Soil Taxonomy for some Calcic great groups and some Calcic subgroups. In addition, the term *free carbonates* is used in several places in the keys to reflect soil materials having finely disseminated carbonates that effervesce in cold, dilute (0.1 M) HCl. See *calcareous class* and *free carbonates.*

Calcium carbonate equivalent (CCE)—The amount of carbonates in the soil (weight percent) relative to the dry total sample weight. A monometric test involving treatment with HCl and measurement of evolved CO_2 is used. The CCE is used in the definitions of calcic horizon and in several taxonomic classes, from suborder to family. Calcium carbonate affects fertility, erosion susceptibility, and water-holding capacity.

Cation-exchange activity class—A family class term used in selected taxa of mineral soils to express their relative level of clay activity. It is used primarily for soils with a combination of mixed or siliceous mineralogy and a clayey or loamy particle-size class. It is calculated as the ratio of CEC-7 to the weight percent of silicate clay and reported as a dimensionless weighted average for the control section. Four classes are used: superactive, active, semiactive, and subactive. It is intended to improve interpretations related to the nutrient-holding capacity and the general level of clay activity for the soils assigned this class. See *apparent cation-exchange capacity.*

Cation-exchange capacity (CEC)—A measure of the negatively charged exchange sites within the soil, expressed as centimoles of charge per kilogram of soil. It is the sum of exchangeable bases measured at pH 7.0 or 8.2. The total CEC may vary significantly depending on the pH at the time of measurement due to the amount of pH-dependent charge present (generally more important in highly weathered soil materials).

Complex (map unit)—A map unit consisting of two or more dissimilar major components that may occur in a regularly repeating pattern or in an unpredictable pattern. The major components of a complex cannot be mapped separately at a scale of about 1:24,000. Each major component is normally present in each delineation, but its proportions may vary appreciably from one delineation to another. See *component*.

Component (map unit)—Within the context of a map unit, a component is an entity that can be delineated at some scale. It is commonly a soil, but it may be a miscellaneous area. Components consisting of soil are named for a soil series or a higher taxonomic class. Those that are miscellaneous areas have an appropriate name, such as Rock outcrop or Urban land. In either case, each component that makes up a map unit can be identified on the ground as well as delineated separately at a sufficiently large scale.

Concretion—A kind of concentration within a soil horizon consisting of a chemical compound, such as calcium carbonate, that is cemented to form a discreet body. It can be removed from the soil intact and exhibits crude internal organization. The organization within a broken specimen is evident to the naked eye as concentric layers around a point, line, or plane. Concretions are similar to nodules, but nodules do not have evident internal organization. In Soil Taxonomy, calcium carbonate concretions are considered a kind of *identifiable secondary carbonate*. In some cases, it is difficult to determine whether the concretions in a horizon formed in place or formed elsewhere and were transported to their current location. A color change gradation from the matrix to the concretion that is not sharp may suggest in situ formation. Soil horizons containing concretions (or nodules) are indicated with the suffix "c" (e.g., Bsc, Bgc). See *nodule*.

Consociation (map unit)—A map unit dominated by a single soil component (or miscellaneous area). Most of the remainder of the delineation consists of soil so similar to the named soil that major interpretations are not affected significantly. The total amount of dissimilar minor components generally does not exceed about 15 percent if they are limiting and 25 percent if they are nonlimiting.

Diagnostic horizon—A specific kind of soil horizon used to identify classes in Soil Taxonomy. Each diagnostic horizon has a definition describing key characteristics of its physical and chemical properties. The definition consists of a few observable and measurable properties that together reflect major genetic processes. Grouping soils with similar diagnostic horizons therefore results in the grouping of soils formed by similar pedogenesis. The use of diagnostic horizons has been adopted by several other soil classification systems worldwide. Diagnostic horizons are not equivalent to genetic horizons. For example, a genetic horizon labeled "Bt" is not necessarily an argillic horizon. Somewhat surprisingly, the term "diagnostic horizon" is not formally defined anywhere in *Soil Taxonomy.*

Electrical conductivity (EC)—Conductivity of electricity (dS m^{-1}) through a saturated paste extract (or 1:5 soil–water mixture) for subaqueous soil analysis. It is used to infer soluble salt levels in salt-affected soils. The EC is a criterion for the salic diagnostic horizon. It is also used to differentiate freshwater and brackish water subaqueous soil classes. Salt content affects plant growth and plant-available water.

Eluvial horizon—A horizon formed primarily as a result of pedogenic removal of soil constituents such as clay. It includes diagnostic horizons such as ochric epipedon and cambic horizons formed at least partially due to eluviation. Many eluvial horizons are designated by master horizon symbol "E" in profile descriptions. See *eluviation.*

Eluviation.—An important pedogenic process characterized by the loss of materials such as clay, and other constituents from the soil. Examples of morphological features resulting from eluvial processes are skeletans, and clay depletions. Eluviation has also been important in the formation of many cambic horizons. See *eluvial horizon.*

Epipedon—A pedogenic horizon formed at or near the soil surface. Pedogenesis is evidenced by the loss of most rock structure, darkening by organic matter, and/or eluviation. Eight diagnostic epipedons are defined in Soil Taxonomy and used to classify soils at multiple taxonomic levels.

Exchangeable sodium percentage (ESP)—Exchangeable sodium as a percentage of cation-exchange activity (at pH 7). The ESP is used as a criterion for the natric diagnostic horizon and several sodium-related great groups and subgroups. Excessive levels of sodium can inhibit plant growth and adversely affect soil structure and hydraulic conductivity. Horizons with accumulations of exchangeable sodium are commonly designated by the horizon suffix "n" (e.g., Btn).

Family—The category in Soil Taxonomy between the subgroup and the series (the 5th level). Its intended use is for grouping the soils within a subgroup that have similar physical and chemical properties. These properties include capacity factors (such as soil texture, mineralogy, and depth) as well as intensity factors (such as soil temperature, reaction, and cation-exchange activity). Soil families are generally defined by properties that are considered important for the use and management of the soil, such as those related to agronomic or engineering purposes. Less emphasis is placed on factors reflecting genetic processes at the family level compared to the higher categories.

Field capacity—The content of water remaining in a soil soon after having been soaked by rain or irrigation and after free drainage becomes negligible, generally 2 to 3 days, in sandy soils even after few hours. It is expressed quantitatively on a mass or volume basis. It is roughly equivalent to the "very moist" to "nonsatiated" wet soil moisture state classes.

Fine-earth fraction—The part of the soil consisting of mineral particles less than 2 mm in diameter (sand, silt, and clay). In the laboratory, whole-soil samples are treated to remove organic matter and then dried, crushed, and sieved to remove rock fragments so that just the fine-earth fraction remains. The weight percents of sand, silt, and clay of the fine-earth fraction are used to determine texture class as depicted in the USDA textural triangle. Many chemical and physical properties used in determining taxonomic placement are based on analysis of just the fine-earth fraction of the soil. See *whole soil.*

Free carbonates—Unbound fine particles of soil carbonates disseminated throughout the soil matrix (as indicated by effervescence when treated with cold, dilute HCl). The term "free carbonates" is used as a diagnostic soil characteristic for mineral soils described in Soil Taxonomy. Although the particles are generally composed of calcium carbonate, they may also be sodium carbonate or magnesium carbonate. Free carbonates may be a precipitate dispersed throughout the soil matrix, or they may have been present in the original parent material. Their presence therefore does not automatically imply pedogenesis. In some situations, their presence or absence in a soil horizon may infer leaching, or lack of leaching. Presence (or absence) of free carbonates is used as a criterion for several taxa and also as part of the definition of the isotic mineralogy class. See *calcareous soil* and *identifiable secondary carbonates.*

Great group—The category in Soil Taxonomy between the suborder and subgroup (the 3rd level). Its intended use is for grouping soils within a suborder that have similar properties that affect soil development in addition to those identified at the suborder level. These properties are commonly static, such as the presence of a particular diagnostic horizon, soil temperature or moisture conditions, base status, and pattern of saturation.

Gypseous (particle-size substitute class)—A family class term used instead of a particle-size class for soils with a high content of the mineral gypsum. The soil contains 40 percent or more, by weight, gypsum in the part consisting of particles less than 20 mm in diameter (e.g., medium gravel and finer). Three gypseous classes are defined. From coarsest to finest, they are "gypseous-skeletal" for those with 35 percent or more rock fragment content, "coarse-gypseous" for those with 50 percent or more particles ranging from 0.1 to 2.0 mm, and "fine-gypseous" for those with more than 50 percent particles less than 0.1 mm. Gypsum is relatively soluble in water and easily leached from the soil. It is therefore found primarily in arid areas where little leaching takes place. Soils with high gypsum contents are susceptible to dissolution and subsidence if subject to repeated water run-on from adjacent built-up areas or from irrigation. See *particle-size substitute class*.

Horizon—A term used for a layer in the soil in which some or all soil properties have resulted from pedogenic processes. The horizon is more or less parallel to the soil surface and can be distinguished from adjoining horizons or layers by differences in one or more of its properties. See *layer* and *horizon nomenclature*.

Identifiable secondary carbonates—A diagnostic soil characteristic of mineral soils consisting of translocated calcium carbonate that has precipitated from the soil solution and was not simply inherited from a soil parent material. It is used in the definition of the calcic diagnostic horizon and for several taxa that have secondary carbonates. Identifiable secondary carbonates are the result of pedogenic processes. They may be soft masses, filaments, cemented nodules or concretions, or coatings within pores or on ped or fragment surfaces, or they may completely engulf a horizon. See *free carbonates*.

Illuvial horizon—A soil horizon that formed primarily as the result of pedogenic accumulation of soil constituents such as clay, soluble salts, and carbonates. The majority of diagnostic horizons are formed by the illuvial process. They include argillic, calcic, gypsic, salic, and natric subsoil horizons. See *illuviation*.

Illuviation.—An important pedogenic process resulting in the movement and subsequent accumulation of materials such as clay, calcium carbonate, gypsum, in a soil horizon. Morphological features resulting from illuvial processes include clay films, carbonate masses, argillic horizons, calcic and petrocalcic horizons.

Infiltration—The movement of water across the interface between air and soil into the soil. The rate of infiltration is highly dependent on near-surface soil conditions and the water state immediately preceding the application of water.

It is also influenced significantly by vegetative cover, cropping systems, and soil management practices and may change seasonally. See *saturated hydraulic conductivity*.

Inherent soil properties—Properties that are little affected by changes in land use and management. Examples include texture, color, and mineralogy. Inherent soil properties are resistant to change and tend to define the basic limitations and potentials for soil use and management.

Layer—A zone within a soil profile for which some or all soil properties are thought to be inherited from the parent material and not the result of pedogenesis. In this case, "layer" is used rather than "horizon." The term "layer" may also be used when no judgement is made as to the genetic origin of the zone. See *horizon*.

Leptic (subgroup)—A taxonomic term used at the subgroup level (an extragrade) in some soil great groups (not all) to denote soils exhibiting relatively shallow profile development. In Kuwait Leptic class is used for Haplogypsids, a gypsic horizon is present at a depth of less than 18 cm.

Lithic contact—A diagnostic soil characteristic consisting of the two-dimensional horizontal plane separating the soil profile above from the strongly (or more) cemented, laterally continuous bedrock below. Cracks are separated by 10 cm or more. The degree of cementation is such that digging with hand tools is impractical. The presence of a lithic contact, depending on its depth, is a diagnostic criterion at the subgroup level of Soil Taxonomy. Where it is too deep for the Lithic subgroup classification, the lithic contact is commonly used to define soil series. The series control section extends to, but not below, the depth of the lithic contact. In a soil profile description, the lithic contact generally coincides with the upper boundary of a layer designated by the master horizon symbol "R." See *paralithic contact*.

Map unit (soil)—A collection of areas with soil components or miscellaneous areas that are both defined and named the same. Each map unit differs in some respect from all others in a survey area and is uniquely identified by a symbol on a soil map. Each individual area (polygon) on the map is a "delineation." See *component*.

Mineral soil material—Soil material with properties that are dominated by the mineral component of the soil rather than the organic part. Mineral soil material contains roughly less than about 35 percent organic matter (or less than about 20 percent organic carbon).

Mineral soil surface—The datum (a horizontal plane) commonly specified in Soil Taxonomy where measurements start for depth or thickness in profiles of mineral soils, for use in evaluating criteria in the taxonomic keys. It is the top of the first layer comprised of mineral soil materials.

Miscellaneous area—A special kind of map unit component that does not meet the definition of soil as given in Soil Taxonomy. These areas are incapable of supporting vegetation without major reclamation efforts. Examples include Rock outcrop, Oil-waste land, and Urban land.

Mottles—Spots or blotches of different color or shades of color interspersed more or less randomly with the dominant color of the soil matrix.

Munsell color—A widely accepted color designation system using notations for hue, value, and chroma (e.g., 10YR 6/4).

Nodule—A kind of concentration within a soil horizon consisting of a chemical compound, such as calcium carbonate, that is cemented to form a discreet body. It can be removed from the soil intact and has no evident orderly internal organization. Durinodes and iron/manganese nodules are specific kinds of nodules recognized as diagnostic features in Soil Taxonomy. Calcium carbonate nodules are considered a kind of *identifiable secondary carbonate.* Nodules are similar to concretions but concretions have evident internal organization. In some cases, it is difficult to determine whether the nodules in a horizon formed in place or formed elsewhere and were transported to their current location. If the color change gradation from the matrix to the nodule is not sharp, in situ formation is likely. Soil horizons containing nodules (or concretions) are indicated with the suffix "c" (e.g., Bsc, Bgc). See *concretion.*

Order—The highest (and most generalized) level in Soil Taxonomy. All of the world's soils can be grouped into one of the 12 orders. The intended use is for grouping soils based on the presence (or absence) of diagnostic horizons or features, which reflect major soil-forming processes that mostly operate over large physiographic regions. The distinctions between orders are based on the evidence of processes that are dominant forces shaping the character of the soil profile.

Organic carbon—Soil carbon originating from organic sources (e.g., decaying plant or animal tissue) rather than inorganic sources (such as calcium carbonate from limestone). It is determined in the laboratory by using dry combustion to measure total carbon and then subtracting inorganic carbon. It is commonly reported as percent by weight. Organic matter generally has high CEC and therefore contributes to nutrient cycling and retention. It has beneficial effects on physical properties such as soil structure, aggregate stability, water-holding capacity, bulk density, and permeability. Organic carbon (as organic matter) is known to be an important soil quality/health indicator. Soil organic carbon cycling is important for terrestrial carbon cycle studies. To convert organic carbon values to organic matter, multiply by the Van Bemmelen factor of 1.724.

Paralithic contact—A diagnostic soil characteristic consisting of a two-dimensional horizontal plane separating the soil profile above from the cemented, laterally continuous soft bedrock below. The cementation class of the bedrock is very weakly coherent to moderately coherent. The bedrock has either been partly weathered or is naturally only weakly consolidated. The underlying paralithic material is mostly continuous, meaning cracks allowing penetration by roots are separated by 10 cm or more. The presence of a paralithic contact, depending on its depth, is a diagnostic criterion at the family or series level of Soil Taxonomy; where it is too deep for family criteria, it is commonly used to define soil series. The series control section is allowed to extend to 25 cm below the paralithic contact. In a soil profile description the layer below the paralithic contact is typically designated by the horizon symbol "Cr." See *paralithic materials* and *lithic contact.*

Parent material—The original unconsolidated mineral or organic material from which the soil formed. Once pedogenic processes act upon and alter the material making up the solum of the soil, the nature of the parent material can only be inferred by the properties of the genetic horizons. The parent material may have been essentially the same as the unaltered soil material currently lying below the solum, or the material from which the soil formed may have all been subject to pedogenic processes and can no longer be observed at the site. The latter is generally the case for soils with one or more lithologic discontinuities. Specific kinds of parent material (other than organic soil material) are not used as criteria in the taxonomic keys for classifying soils. Largely unaltered mineral soil material is generally designated by the horizon symbol "C."

Particle-size class (taxonomic)—One of several class terms that together make up the family level of Soil Taxonomy. The terms are designed to characterize the grain-size composition of the whole soil, including both the fine-earth material and the rock and pararock fragments, as well as some human artifacts. They do not include organic matter and salts more soluble than gypsum. The definitions of the classes represent a sort of compromise between particle-size groupings commonly used in pedology and those commonly used in engineering. Exceptions to rules include treating clay-size carbonates as silt and allowing very fine sand particles to be considered as either sand or silt, depending on the texture class of the fine-earth material.

Particle-size distribution—The proportions of mineral particles of various sizes making up the soil. Depending on the purpose, the distribution of particles considered may be limited to just the fine-earth material (i.e., sand, silt, and clay) or it may be for the whole soil, thus including the fine-earth material as well as rock and pararock fragments. The relative distribution of the various sizes of particles can be used to categorize the soil material using various systems, such as the USDA texture classes (sandy loam, loam, etc.) and AASHTO (A-2, A-3, etc.) and Unified (SP, GP, etc.) engineering textural classes, as well as other systems.

Particle-size substitute class—A family class term used instead of a particle-size class for soil materials that either cannot be dispersed effectively by standard laboratory methods or that contain too little fine-earth material for reliable and useful particle-size analysis. Substitute class terms are used for soils with high amounts of gypsum. See *particle-size class*.

Ped—A single soil structural unit that formed as a result of pedogenic processes. The ped surface persists through successive cycles of wetting and drying. Structural units not caused by pedogenesis are called clods. See *soil structure*.

Pedon—A three-dimensional body of soil that encompasses sufficient area (roughly 1 to 10 m^2) and depth (up to 200 cm) to enable description of the internal arrangement of horizons and collection of representative samples for laboratory analysis. The pedon is the soil individual classified in Soil Taxonomy. Descriptions of soils are commonly referred to as "pedon descriptions"; in practice, however, it is more accurate to call them "profile descriptions" because most are a description of a two-dimensional face of an exposed soil, as in a pit or road cut, rather than the result of the dissection of a three-dimensional volume of soil. See *soil profile*.

Quality assurance (soil survey)—The process of providing technical standards and guidelines, oversight and review, and training to ensure that soil survey products meet defined standards. Responsibility for assuring quality in soil survey products rests primarily with soil scientists in supervisory and administrative positions. Examples of quality assurance activities include carrying out periodic reviews of completed work, conducting training, developing technical notes to clarify the application of standards, reviewing the technical content and accuracy of manuscripts and databases, and similar actions. See *quality control.*

Quality control (soil survey)—The routine implementation of the collective set of standards and procedures adopted in soil survey. Controlling quality involves the competent application of soil survey standards and procedures on an ongoing basis by those directly performing soil survey tasks, whether in the field, in the laboratory, on cartographic or editorial staffs, etc. It also includes providing direct review and inspection, direction, and coordination of soil survey production activities by supervisory soil scientists. These supervisory scientists do this by ensuring that staff members have the equipment and training necessary to complete their tasks and that their soil survey products meet the defined standards for content, accuracy, and precision. When deficiencies are discovered they are quickly remedied, be it through training, acquisition of proper equipment, mentoring, etc. See *quality assurance.*

Ratio, 1500 kPa water content to clay—A calculation that is used to define the relationship between water content at 1500 kPa tension (wilting point) and measured clay percent. In general, if a sample is effectively dispersed during standard pretreatment in the laboratory, the ratio of 1500 kPa water to clay dominantly falls between 0.25 and 0.6, most commonly around 0.4. Values of 0.6 or higher suggest poor dispersion, and the particle-size distribution results are considered unreliable. Low values tend to result from the dominance of low-activity clays, gypsum, or carbonates. High values are common for soils. The rules of application for family particle-size class require that clay be calculated with the formula: clay = 2.5 (1500 kPa water -% OC) when the calculated ratio of 1500 kPa water to clay is 0.25 or less, or 0.6 or more. In addition, a ratio of 0.6 or more is one of the diagnostic criteria for the isotic mineralogy class.

Ratio, CEC-7 to clay—A calculation that is used to define the relationship between measured cation-exchange capacity (by ammonium acetate at pH 7) and silicate clay. It can be used to generally infer the kinds of clay minerals present. Larger values indicate the presence of clay minerals with a higher density of cation-exchange sites, such as illite or smectite. Smaller values suggest the presence of low-activity clay minerals, such as kaolinite, halloysite, or gibbsite. The ratio of CEC-7 to clay is used with some loamy or clayey soils with mixed or siliceous mineralogy to calculate CEC activity classes for soil family placement in Soil Taxonomy. This allows one to make more precise interpretations about cation retention and chemical movement in these soils than could be made from texture and mineralogy class alone. See *cation-exchange activity class.*

Rock fragments—Discreet mineral fragments 2 mm or more in size that have a strongly coherent or higher rupture-resistance class (i.e., gravel, cobbles, stones, etc.). Their origin may be geologic (e.g., pieces of granitic rock inherited from the parent material) or pedogenic (e.g., unattached pieces dislodged from a petrocalcic horizon).

Root-limiting layers—Specific kinds of layers or horizons that are root restrictive in nature. Unless otherwise indicated in Soil Taxonomy, the diagnostic horizons or features considered to be root-limiting layers are: duripans; fragipans; petrocalcic, petrogypsic, or a manufactured layer; and densic, lithic, paralithic, and petroferric contacts. Embodied within the definitions of most of these features is the requirement that roots entering these materials are limited to cracks or fractures no closer than 10 cm apart. It should be noted that the concept of root-limiting layers in Soil Taxonomy is strictly as a physical impediment to roots. The top of a root-limiting layer is the lower depth limit for the application of most soil family criteria and for determining thickness or depth in some of the other definitions used in Soil Taxonomy.

Saturated hydraulic conductivity—A quantitative measure of a saturated soil's ability to transmit water when subjected to a hydraulic gradient; commonly abbreviated as "Ksat." It is formally defined as the proportionality factor that relates water flow rate to hydraulic gradient in Darcy's equation. Although it is commonly reported in units of length/time (meters/second, inches/hour, etc.), this form of expression only applies in the special case when the hydraulic gradient is unity (numerically equal to flux). Ksat expressed by length/time values is useful for comparison and general ranking purposes, but does not represent the actual rate of water movement at any particular site. Saturated hydraulic conductivity is a property of the soil that governs the ease with which pores of a saturated soil permit water movement. Ksat measurements in the field are subject to large variation within relatively small areas. Therefore, it is important to take several measurements at a site to account for this natural variability. Estimates for general Ksat class placement are sometimes made based on texture and bulk density.

Skeletal—A term used in conjunction with some particle-size or substitute class terms at the family level for soils having 35 to 90 percent, by volume, rock fragments (e.g., sandy-skeletal). Artifacts that are 2 mm or more in size and both persistent and cohesive are included with rock fragments for this determination.

Soil—For purposes of soil survey and soil classification, soil is defined as a natural three-dimensional body comprised of solids (minerals and organic matter), liquid, and gases that occurs on the land surface (including shallow water areas), occupies space, and is characterized by one or both of the following: (1) horizons, or layers, that are distinguishable from the initial material as a result of additions, losses, transfers, and transformations of energy and matter; *or* (2) the ability to support rooted plants in a natural environment. For soil survey purposes, anything that is not soil is considered a *miscellaneous area.*

Soil genesis—The origin and development of the soil as determined by the combined action of the soil-forming factors (parent material, relief, climate, organisms, and time). The original parent material is altered, and horizons and other morphological features develop in a soil profile. Pedology combines the fields of soil physics, chemistry, mineralogy, and biology to study the genesis of soils. Soil Taxonomy, through the use of diagnostic horizons and features, is designed to reflect current understanding of soil genesis but does not directly include theories of soil genesis in the criteria for classifying soils.

Soil morphology—The physical composition of a soil profile as expressed by the kinds, thickness, and arrangement of soil horizons and by other internal characteristics (such as texture, color, structure, consistence, pedogenic accumulations (or losses) of substances and porosity). A major purpose of Soil Taxonomy is to group soils that exhibit similar morphology.

Soil profile—A more or less two-dimensional vertical cut through the soil on a plane at right angles to the soil surface. It is an exposure used for describing the kinds and arrangement of horizons making up the soil. See *pedon.*

Soil separates—The grouping, by size, of mineral soil particles that are less than 2.0 mm in equivalent diameter into discreet classes, which range from clay to very coarse sand. In soil survey the most commonly used set of classes is the USDA system. Other systems, such as the Unified and AASHTO, are also recognized for specific uses. Laboratory measurements of the various soil separates making up a sample use a combination of differential settling in a water column (e.g., pipette method) and dry sieving. See *texture class.*

Soil series—The lowest level in Soil Taxonomy. All the soils of a series have horizons that are similar in composition, thickness, and arrangement. A soil series is a conceptual class defined to represent natural bodies of soils (polypedons) on the landscape. Soil series are defined by properties within the series control section as defined in Soil Taxonomy.

Soil structure—The arrangement of primary soil particles into natural aggregates (peds). The cohesive forces holding individual peds together is greater than the adhesive forces among peds. As a result, the surfaces of individual peds persist through successive wetting and drying cycles and the soil mass ruptures along these predetermined planes. Some soil horizons are structureless and described as either single grain or massive. The development of structure is an important consequence of soil formation. See *ped.*

Soil surface—The horizontal plane consisting of the upper boundary of the soil with overlying nonsoil such as air or water. The horizon at the soil surface may be either mineral or organic. It does not include undecomposed plant materials. Some depth or thickness measurements used for classifying soils specify measurements that begins at the soil surface. See *mineral soil surface.*

Soluble salts—Salts that are readily removed from the soil by water alone. In general, they are salts containing sodium, calcium, magnesium, potassium or bicarbonate chloride, and sulfate. They include the salts contributing to the overall electrical conductivity measured in a saturation extract (ECe). Soils containing appreciable soluble salts dominantly occur in semiarid or arid areas where normal leaching from precipitation is not sufficient to remove the salts from the profile. Diagnostic horizons that commonly contain accumulations of soluble salts include anhydritic, salic, natric, gypsic, and calcic horizons. Soil profile descriptions use horizon suffix symbols (such as z, n, y, or k) to indicate salt accumulation (e.g., Bz, Btn, Bky).

Solum—The part of a soil profile that formed by pedogenic processes; commonly made up of horizons designated A, E, and B, but not including the underlying C or R horizon. The solum consists of a sequence of soil horizons that are related through the same cycle of pedogenic development. It includes horizons below a discontinuity if the lower horizons acquired some properties by currently active soil-forming processes. It therefore does not include a buried soil. The solum is not necessarily confined to the zone of major biological activity. Its genetic horizons may be expressed faintly or prominently. A solum does not have a defined maximum or minimum thickness. The identification of the exact limit of the solum in the soil is somewhat subjective. Although used to communicate information about soil profile development in a general way, the solum is purposely not used as a criterion anywhere in Soil Taxonomy due to its subjective interpretation. Many soil series descriptions, however, include a statement about solum thickness in their range in characteristics.

Strongly contrasting particle-size classes—A set of terms used to classify soils that have two layers within the particle-size control section made up of particle-size classes (or substitute classes) that contrast strongly with each other (e.g., sandy over clayey). Each layer must be at least 12.5 cm thick, and the transition zone between them must be less than 12.5 cm thick. Water movement and rooting patterns are significantly affected in soils with abrupt changes in texture.

Subgroup—The category in Soil Taxonomy between the great group and family (the 4th level). Its intended use is for grouping soils within a great group that have properties shared by similar soils in other categories or that have unique properties not recognized in any other class.

Suborder—The category in Soil Taxonomy between the order and great group (the 2nd level). Its intended use is for grouping soils within an order that have some key soil property that exerts a controlling influence over currently active soil-forming processes. This key property commonly consists of dynamic climatic controls such as the soil moisture or temperature regime, but in some cases includes the presence of specific diagnostic horizons, sandy textures that resist pedological development, stratified layers indicating relatively rapid rates of soil accretion from deposition, or significant accumulation of organic matter in the subsoil.

Surface mantle of new soil material—A technical phrase used to describe a particular kind of recent surface deposit. The deposit is composed of mineral soil material at least 50 cm thick, has a zone at least 7.5 cm thick at its base that fails the criteria for any diagnostic horizon, and is underlain by an older soil with a sequence of one or more pedogenically developed horizons. The presence of a surface mantle of new soil material is a requirement for recognizing a buried soil for purposes of soil classification. Where a surface mantle of new soil material overlies a buried soil, the soil profile essentially contains two soils, a newer one overlying an older one. Special care is needed to correctly classify such soils. Specific rules are provided in Chap. 4 of the *Keys to Soil Taxonomy*. See *buried soil* and *buried horizon*

Texture class—The proportions (by weight) of various size classes of soil separates (sand, silt, and clay) making up a soil sample. Definitions for 12 classes are provided in the *Soil Survey Manual* and depicted in the USDA textural triangle (sand, loamy sand, sandy loam, loam, etc.). The classes of sand, loamy sand, and sandy loam can each be further subdivided into four subclasses based on dominant sand size (very fine sand, fine sand, sand, coarse sand, etc.). Texture class modifiers for samples containing material > 2.0 mm are used as appropriate (e.g., gravelly loam). In addition to the detailed classes described above, the texture classes can be grouped into generalized classes, such as sandy, loamy, and clayey, as described in the *Soil Survey Manual*. In the field, soil texture is estimated by tactile tests, such as feeling a moist sample and ribboning. Laboratory measurements of the various soil separates making up a sample use a combination of differential settling in a water column (e.g., pipette method) and dry sieving. Texture class is not the same as the particle-size class used to define soil families in Soil Taxonomy. See *apparent field texture*, *soil separates*, *particle-size class (taxonomic)*, and *particle-size distribution*.

Transitional horizon—A soil horizon whose dominant properties reflect one type of master horizon but which has subordinate properties characteristic of another type of master horizon. Two capital letters are used for the horizon designation, such as "AB" or "BC," with the first symbol being the dominant of the two horizon types. Transitional horizons are commonly recognized in a position between two kinds of horizons where the horizon characteristics are transitioning from one kind of horizon to another, but they can also be recognized when one of the presumed kinds of horizons is no longer present (e.g., B-CB-R). See *combination horizon* and *horizon nomenclature*.

Typic (subgroup)—The subgroup that contains all the soils that do not meet the criteria defined for the other subgroups within a great group. These soils are not necessarily the most extensive soils nor do they necessarily represent the central concept of the great group. In all cases however, the Typic subgroup is the last subgroup listed in the key to subgroups within that particular great group.

Typical pedon—A specific pedon description chosen to represent a map unit component as it occurs in a soil survey area. It is used in the correlation process to classify, name, and interpret the component. The pedon description, along with information about the soil's overall range in characteristics, landscape setting, and other pertinent information, is included in the soil survey publication and/or database. The typical pedon is not a composite description based on a collection of pedon descriptions, but rather it is a real pedon with a physical location that can be revisited (also called a type location).

Undifferentiated group (map unit)—A map unit consisting of two or more major soil components that are not consistently associated geographically and therefore do not always occur together in the same map unit delineation. These components are included as members of the same map unit because use and management considerations for each soil are the same or very similar for common uses. Generally, they are included together because some common feature outside of the soil itself, such as very steep slopes, extreme stoniness, or frequent flooding of long duration, is an overriding consideration for use and management.

Weighted average—A simple mathematical calculation used to determine the average amount of a soil constituent (e.g., percent clay) within a specified part of the soil (commonly the particle-size control section). It is useful because soils are typically described and sampled in the field according to individual genetic horizons and layers, which rarely coincide exactly with the control section depths specified in Soil Taxonomy. A weighted average is calculated by summing the product of the measured value of the constituent times thickness for each layer within the control section and dividing by the total thickness of the control section. It is a way to simulate the mixing of multiple samples and allows one to obtain a single average value for the constituent of interest. Many criteria in Soil Taxonomy are based on a weighted average value.

Whole soil—Soil material that includes all particle-size fractions, up to and including boulders, that have maximum horizontal dimensions less than those of the pedon. Many chemical and physical criteria are specifically determined only for the fine-earth (<2.0 mm) fraction of the soil. Some, however, particularly rock fragment content, are determined on a whole soil basis. See *fine-earth fraction*.

Reference

Ditzler C (2017) A glossary of terms used in soil survey and soil classification Including definitions and brief commentary. Prepared for NRC, USDA https://www.nrcs.usda.gov/wps/PA_NRCSConsumption/download?cid=nrcseprd1806427&ext=pdf

Author Index

S. A. Shahid and S. A. S. Omar, *Kuwait Soil Taxonomy*,
https://doi.org/10.1007/978-3-030-95297-6

Subject Index

S. A. Shahid and S. A. S. Omar, *Kuwait Soil Taxonomy*,
https://doi.org/10.1007/978-3-030-95297-6

T

U

W

X

GPSR Compliance

The European Union's (EU) General Product Safety Regulation (GPSR) is a set of rules that requires consumer products to be safe and our obligations to ensure this.

If you have any concerns about our products, you can contact us on ProductSafety@springernature.com

In case Publisher is established outside the EU, the EU authorized representative is:

Springer Nature Customer Service Center GmbH
Europaplatz 3
69115 Heidelberg, Germany

Batch number: 10614849

Printed by Printforce, the Netherlands